JAMES HUTTON

THE GENIUS OF TIME

James Hutton

The Genius of Time

RAY PERMAN

BIRLINN

First published in 2022 by
Birlinn Ltd
West Newington House
10 Newington Road
Edinburgh
EH9 1QS

www.birlinn.co.uk

ISBN 978 1 78027 785 1

British Library Cataloguing in Publication Data
A catalogue record for this book is available from the British Library.

Typeset by Hewer Text UK Ltd, Edinburgh

Printed and bound by Clays Ltd, Elcograf S.p.A

Contents

Preface

JAMES HUTTON'S ABILITY to step outside the constraints of time enabled him to change our thinking with his revolutionary *Theory of the Earth*. His rejection of the biblical narrative, which held the Earth to be little over 6,000 years old, led him to be attacked and his concept of infinite time described as an abyss, a cold, dark, Godless void. But looking into the chasm held no terror for him. He saw clearly a simple and repeated pattern of destruction and renewal, which had shaped the surface of the world over unimaginable ages and continued to do so.

Hutton could read a stone like a book. A cliff face was a library, and as he and two friends gazed on Siccar Point, in Berwickshire on the east coast of Scotland on a calm spring day in 1788, he was able to transport them back to tell them the story of the globe in the simplest language. 'The mind seemed to grow giddy by looking so far into the abyss of time,' one of them remembered, 'and while we listened with earnestness and admiration to the philosopher who was now unfolding to us the order and series of these wonderful events, we became sensible how much farther reason may sometimes go than imagination can venture to follow.'

There are already at least five biographies of Hutton, beginning in 1805 with an account of his life and works by John Playfair, one of the two friends who was with him and looked in wonder at the rock strata. Playfair knew him and became a disciple, yet despite his account Hutton remains an enigma. The *Theory of the Earth* arguably changed our thinking as much as the books of his contemporaries Adam Smith and David Hume, but he remains much less known than they are. There are big gaps in his life story, important periods when we have no direct evidence of what he was doing.

Apart from his published works we have very few of his own papers – a handful of letters either to him or from him to others, no diaries or journals, no books of accounts, not even a will.

Playfair tells us that he 'wrote a great deal and has left behind him an incredible quantity of manuscript'. Yet very little of this has survived and the fate of the rest is a mystery. Several scholars have searched in the hope of finding a vast cache of Hutton papers in some dusty attic, but in vain. We also know that he was a voracious reader, yet his library also disappeared – and with it clues to some of the scientists, explorers and theorists who might have influenced him. We have more of an idea about what happened to his extensive collection of rocks and fossils, but it too has become so dispersed that it is impossible now to say with certainty that any sample now in a museum or collection once belonged to him.

The picture of Hutton I have tried to paint has been based on as much evidence as I can find of what he was doing and what he was thinking, but I have worked under two handicaps: the first is the lack of documentary evidence mentioned above. Where I have had to assume actions or to interpolate to fill the gaps between known facts, I have tried to make this explicit in the text. The other hindrance was the global coronavirus pandemic, which began in 2020 and closed libraries and archives, particularly those in Edinburgh, London and Paris. I am grateful to those librarians and archivists who, although often working from home, tried to help with my inquiries nonetheless. My appreciation too to Sir Robert Clerk for permission to quote from the Clerks of Penicuik archives.

My special thanks to Ian Robertson, whose written and telephone French is far superior to mine, and who spent many lockdown hours chasing leads and suggesting new directions for research. He also read and made many helpful comments on the whole manu-script. Professor Stuart Monro provided geological knowledge which I lacked and also made perceptive comments on the draft book. Professor Alan Werritty, who at the time of writing, was preparing Hutton's unpublished manuscript on agriculture for publication, drew my attention to important aspects which I had missed and made helpful comments on other aspects of his life. I am grateful to Jeanne Donovan for sharing her research on Hutton's family. My wife, Fay Young, has put up with James Hutton as a lodger in our house for the past two years and tolerated my amateur attempts to point out 'unconformities' during our country walks. Finally, I owe a debt to all the staff of the James Hutton Institute, which I had the privilege of chairing for six years, for opening my eyes to the genius of Hutton himself.

Ray Perman
Edinburgh, July 2022

Chapter 1

He cared more about fossils . . .

TO HIS CLOSEST friends James Hutton was a lifelong bachelor, but when he died they realised they had not known him as well as they thought. The news of his death was given to the steam engine builder James Watt by the physicist and mathematician John Robison. Overcome with grief, it was nearly three months before Joseph Black, who had been closest to Hutton and frequently at his bedside during his final illness, could bring himself to write to Watt, and when he did it was to reveal a shock. Shortly after Hutton's death on 26 March 1797, a man had arrived in Edinburgh claiming to be his son. He was also called James, and Black guessed him to be about 50 years old. He had lived for most of his life in London where he was a clerk in the Post Office, but poor health had obliged him to seek leave of absence and move to the village of Flimby, near Workington in west Cumbria, where one of his sons was apprenticed to a doctor.[1]

The tone of the letter makes it clear that Black and Watt, two of Hutton's most intimate confidents for more than 20 years had no idea he had a son – or that he had ever had a relationship. Although he enjoyed the company of women, he was – or at least appeared to be – unmarried and shared his house in St John's Hill, Edinburgh, with his older sister, Isabella.[2] The man professing to be his son could easily have been taken for a charlatan out to claim a share of Hutton's fortune, but Black was convinced he was genuine.

I have conceived a warm affection for this poor man, he is not unlike the doctor* in person and bald like him but not like him in the face, having more of the features of his mother, he is a

* 'Dr' in the original. Hutton's friends called him 'the doctor', perhaps ironically because although qualified in medicine he never practised.

most worthy modest man, of great assiduity and ability in business but has been so much confined as to be but little acquainted
with the tricks of this world.

The younger Hutton was married with four sons and three daughters, one of whom he brought with him from Cumbria to Edinburgh
to be his nurse and companion. Black thought her well educated
and was so taken with the newly discovered Hutton family that he
later arranged for the son who wished to be a doctor to study at
Edinburgh University and for another son to become an apprentice
to Watt's steam engine business in Birmingham. The younger James
Hutton died in 1802, five years after his father, but his children
were accepted and absorbed into the family. They were treated
kindly by Hutton's friends, Isabella supported them financially, and
when she died left them two farms the family owned. But if James
Hutton senior was their grandfather, who was their grandmother
and why did Hutton keep her identity and their existence a secret
even from his closest companions for nearly 50 years?

The mystery has intrigued Hutton's modern biographers and
researchers and led to many theories – some plausible, others wild
– on the slimmest of evidence. It is part of Hutton's enigma. He is
one of the major figures of the Scottish Enlightenment, whose
insights contradicted the established orthodoxy and changed the
way we think about the Earth and its evolution. He has been called,
with some justification, the father of modern geology. Yet other than
his published works on geology, agriculture, science and philosophy and a brief outline of his life, we know very little about him.

Around 170 letters survive from or to Adam Smith, the famous
economist and another of Hutton's friends, yet we only have a handful of Hutton's letters. No household accounts, business papers,
diaries or memoirs survive, not even a last will and testament. At
Smith's request Hutton and Black, who were his executors, burned
his private papers after his death. The great economist wanted to be
remembered and judged on his considered published works, not his
personal life. Perhaps Hutton was the same. Although for nearly 50
years he was in the centre of the Edinburgh Enlightenment, he stood
apart from it. He was not a public man. He did not teach at the
university as Black did, nor have an official position as Smith did.

He lived through tumultuous times – the occupation of Edinburgh by the Jacobite army in 1745, the American War of Independence, the French Revolution – yet his experiences or views on any of them are unknown to us. So what do we know?

Hutton was a modest man, he lived quietly and unostentatiously. His surviving portraits contrast strongly with those of other Enlightenment figures. The philosopher David Hume was painted twice by Allan Ramsay, once in the sumptuous red and gold court uniform of a diplomat and the second time wearing a brocade waistcoat and a red velvet turban. The depiction of Hutton by Henry Raeburn shows him in a simple brown suit, his hands together in his lap, a benign expression on his face. Several waist-coat buttons are undone. Raeburn does not tidy him up, but paints him as he was – a man careless of his public image. He was described by Robert Louis Stevenson as 'in Quakerish raiment and looking altogether trim and narrow and as if he cared more about fossils than young ladies'[3] – and the fossils are on his table, alongside his books and papers.

In all the images we have of him he is bare-headed. Eighteenth-century gentlemen of standing commonly wore wigs – even Smith and Black, the least vain of men, were always pictured in wigs. Yet Hutton saw no need to hide his baldness. His slim figure testified to a frugal diet, unlike the portly *bon viveur* Hume. His first biographer tells us that he ate sparingly and drank no wine.[4] His physique was a reflection of his lifestyle, combining intellectual activity with physical exertion. The Raeburn portrait was painted when Hutton was in his late fifties or early sixties, a period when he was still under-taking long research trips on foot or on horseback. His modesty lasted after death. Unlike Hume, whose funeral attracted a large crowd despite pouring rain, or Smith, whose monument was designed by Robert Adam, the most famous architect of the age, Hutton was buried quietly. Outside a very small circle of friends, his genius went unrecognised after his death and his grave was unmarked for 150 years. Even now, when he is acknowledged to have revolu-tionised our thinking of how the Earth was formed, there is only a simple wall plaque inscribed with his name.

Yet we should not get the idea that Hutton's unpretentious life made him dour or unfriendly. Quite the reverse, he was gregarious and needed company – the one period of his life when he lived mostly alone, during his first years farming in Berwickshire, he was unhappy. He made friends easily and most of the relationships he formed in his youth remained firm throughout his life. There is evidence that his opinion – on personal as well as scientific matters – was sought and trusted. He loved a party and could liven any gathering with his wit and conversation. Joseph Black trying to cheer a depressed James Watt wrote: 'I wish I could give you a dose now and then of my friend Hutton's company; it would do you a world of good.'[5] The very different characters of Black and Hutton complemented each other. According to Black's biographer: 'Black was serious, but not morose; Hutton playful, but not petulant. The one never cracked a joke, the other never uttered a sarcasm.'[6]

John Playfair, Hutton's friend and follower who was his first biographer, says that he enjoyed the company and conversation of 'accomplished individuals of both sexes' and enlivened any group. 'A brighter tint of gaiety and cheerfulness spread itself over every countenance when the doctor entered the room.'[7] Hutton, in one of his few surviving letters, writes longingly of a dream of a house party: 'a mortal deal of fun we had and very busy we were'.[8]

Despite enjoying social position and education, he had few airs and graces. His letters (and presumably his speech) slipped often into the vernacular – even the vulgar – and Playfair tells us that 'to an ordinary man he appeared to be an ordinary man', although he qualifies this by adding 'possessing a little more spirit and liveliness, perhaps, than is usual to meet with'.[9] Playfair's conclusion gives us a picture of a warm and open man: 'He was upright, candid, and sincere; strongly attached to his friends; ready to sacrifice anything to assist them; humane and charitable. He set no great value on money, or, perhaps, to speak properly, he set on it no more than its true value.'[10]

Indeed, Hutton was generous with both his time and his money. He went out of his way to help the young Silas Neville find suitable lodgings and settle in Edinburgh when he came to study medicine.[11] His cousins John and Andrew Balfour had emigrated to America, but John was killed during the War of Independence when his house was burnt down by irregulars. His widow, Mary Ann Gray, decided to return to Scotland with her three daughters and son Andrew, who was less than a year old. Hutton housed and supported the family, and as the boy grew up acted as a surrogate father to him, taking him for rambles in the King's Park and on Arthur's Seat, the volcanic mountain in the centre of Edinburgh.[12] And Hutton contributed to public causes, £50 to the cost of raising a regiment of the Edinburgh Volunteers to fight in the American war[13] and a similar amount to the fund for new buildings for Edinburgh University.[14]

So why did this generous, open and honest man conceal a relationship and a child for half a century?

Playfair, in his biographical sketch, ignored the son altogether, giving no inkling that Hutton had ever had a sexual encounter, although as a close friend of Joseph Black he surely would have known and possibly met the younger James Hutton. We know that he met some of Hutton's grandchildren. Playfair was not only a friend of the older Hutton, he was also convinced by his ideas and became a promoter and defender of Hutton's *Theory of the Earth*. It is likely that he, like Hutton himself, felt that the great man's reputation should rest on his work, not his domestic life. Early modern biographers of Hutton were also ignorant of the existence of a son, which only came to light with the discovery of the letters between Black, Watt and Robison in 1970.

But the subject has intrigued later Hutton scholars and researchers, some of whom have developed extraordinary theories about who the woman was, where and when Hutton met her and why her identity has remained unknown. They have also used the sudden appearance of the younger James Hutton to make judgements on the older Hutton's character which are very different to those of his contemporaries. Some have taken Black's estimate that the son was about 50 (Black added the caveat 'has the appearance of being older') as meaning that he was born in 1747, when Playfair tells us Hutton was in Edinburgh, aged 21 and studying medicine. One has suggested that Black's description of the young man's face as resembling his mother, meant Black knew Hutton's mistress and had met her either when he was in Edinburgh writing his thesis in the early 1750s, or when he returned to the city in 1766.[15] For this to have been the case, both Hutton and his alleged mistress must have been living a lie, keeping the son concealed (in London?)

while they lived separate and to outward appearances innocent lives in Edinburgh.

Others have made a different allegation: that because Hutton left Edinburgh at the end of 1747 to further his medical studies in first Paris and then Leiden, he was escaping the scandal – getting out of Scotland until memories had faded of the birth of his illegitimate child and presumably leaving the mother to bear the shame alone and fend for herself. One of Hutton's most recent biographers even cites the case of David Hume's father, who did exactly that – fled the country to avoid disgrace.[16] If Hutton did leave Edinburgh to avoid dishonour he would have been callous and a coward. In the Church of Scotland of the day, public humiliation was often the consequence of fornication. The poet Robert Burns, who met Hutton in later life, had to suffer the indignity of the 'cutty stool', being made to stand on a low seat in front of the congregation while his conduct was condemned at length from the pulpit.[17]

The minutes of kirk sessions describe pregnant women or new mothers accused of 'adultery and bastardy' being hauled before the minister and elders and told to name the father. The church had a pecuniary as well as a moral reason for doing so; if the man would not provide for his child, there could be a 'burden on the parish'. Sometimes the women would voluntarily admit before the minister that her child had been 'begat in fornication' and ask for a finding of paternity against a named man. Some men and women even signed written confessions.[18] For a God-fearing family with a place in society like the Huttons, the shame of a public admission would have been humiliating. But the name James Hutton does not appear in any of the surviving records of the time, although we cannot take this as proof that he was not accused. Frustratingly, the minutes of the kirk session at Greyfriars, the parish where the Huttons lived and the church in which they probably worshipped, are missing for the relevant dates.[19]

There have been other theories. One serious researcher has suggested that the mother was a blood relation of Hutton, perhaps his cousin or aunt? This would not necessarily have been a reason for secrecy: marriage between cousins was not encouraged by the Church of Scotland, but neither was it prohibited. But perhaps the woman was already married? Even the author of this theory had her doubts, writing on her scribbled notes 'QE not quite D'.[20]

The digitisation of parish records and their availability online for genealogical research has made it possible to come up with new candidates for the mother and when and where Hutton may have met her. The name 'James Hutton', it transpires, was not uncommon in 18th-century Britain and more than one of them had a son in the middle of the century and named him 'James Hutton'. The online quest to identify the mother has placed her in Edinburgh, Norfolk, Yorkshire and Wiltshire among other places. For example, on only circumstantial evidence, she has been identified as Mary Eidington, born in Aberlady, East Lothian, in 1732, which would have made her 14 or 15 in 1747. Thus, by implication James Hutton not only cynically abandoned the mother and child either before or shortly after the birth, but was also a male predator, preying on a girl six or seven years younger than himself for his own gratification, regardless of the consequences. This would not have been uncommon and the legal age for marriage was 12 for girls and 14 for boys, but it is a picture of Hutton as a young man which goes against everything we know about him in later life as a caring and moral individual.

In their letters to each other, Black and Watt speak warmly of Hutton's kindness and character. We know that he went out of his way to help others. He made long-term relationships and was loyal to his friends. We do not know why he did not live with the mother of his son, but we do know he was a caring father and in touch with his son. Hutton supported his son and his family through their education and in other ways. Black's letter to Watt says that Hutton 'supplied money from time to time [to his son] and to whom he intended some time past to leave a legacy'. He adds: 'his [the son's] salary in the Post Office afforded him a scanty and even insufficient subsistence, but the doctor allowed him to draw upon him when necessity required it, which Mr H did as seldom as possible.'[21] Why they lived apart we do not know, but this is not a picture of a father who had abandoned his son.

To complicate the search for the mysterious mother, Black's estimate of the son's age was wrong. In fact he was two years younger. The Flimby parish register records the date of death of James

Hutton junior as 1 February 1802, and his age as 53.[22] This would put his birth in 1748 or the beginning of 1749. According to Playfair the older Hutton had left Edinburgh by then and was in Paris studying anatomy and chemistry. So could the mother be French? It is possible. Forty years later the poet William Wordsworth fathered a child during his stay in France. But like all the other speculations there is no reliable evidence that Hutton did the same. It is also possible that the child was conceived in Edinburgh before Hutton left, or at any number of places on the journey to Paris. We just don't know.

Playfair may not be right about dates and places. He was a friend and follower of James Hutton, but he did not meet him until 1781, when Hutton was 55, and his biography was written several years after Hutton's death. Although he had access to some documents which have since been lost, he was probably relying on his own memory of Hutton's anecdotes about his life, or on the memories of others, which may not have been reliable. Most of Hutton's friends and collaborators and his surviving sister were old by the time of his death. We know Playfair wrote to James Watt asking for information about a trip he and Hutton had made, only to receive the reply that it was long ago and Watt could not trust his memory.[23] And what of Black's comment that the younger man's features resembled those of his mother? At least one researcher has assumed that this means Black must have known the woman personally.[24] But need this be the case? Black may have simply remarked to the younger Hutton that his face did not resemble his father's and received the reply that he took after his mother.

The identity of this mysterious woman is not the only gap in our knowledge of Hutton's life.

Chapter 2

Not ten days in which I was not flogged

THE EDINBURGH OF the time of James Hutton's birth, although a small city by European standards, was one of the most crowded. It had grown along the narrow spine of the mountain leading upwards and westwards from the Palace of Holyroodhouse to the castle on its precipitous rock. The broad road linking these two landmarks – called the Royal Mile today – had narrow alleys or 'closes' leading off, down the slopes of each side to north and south. Most of the population lived literally on top of each other. Tall narrow tenements (called 'lands'), each housing several families, cramped and shadowed the closes, making them dark and damp. Rich and poor shared the same front door and stairwell – the better-off families on the lower floors, closer to the water pumps and with a faster escape in case of fire, which was a frequent hazard. Poor families lived higher up. All social classes threw their waste – including human waste – into the street. Edinburgh, infamous for its filth and its smell, had no proper sanitation and was short of clean drinking water.

The Hutton family was modestly prosperous. James' grandfather, William was admitted to the Merchant Company of Edinburgh in 1688, at the age of 22, making him a 'burgess' and a 'guild brother' – superior social status in the commercial life of the city. He was elected by his fellows to serve on the town council and also became one of the two treasurers of Trinity Hospital, an alms house administered by the council. He is described in official documents as a merchant – a catch-all name which covered everyone from a wealthy international trader to a pedlar who rarely ventured outside the town, according to one historian[1] – but from archive sources we can deduce that he was a tailor. In 1702 the town council paid him 600 pounds Scots (£50) for lining cloth for uniforms of the Edinburgh company of grenadiers.[2] In 1718 he was sending an 'acompt to My Lady Cringletie' for a riding gown, with 'silk, silver, buchram and lining to the body and sleeves'.[3]

Poll tax and hearth tax records for early 18th-century Edinburgh have not survived, but the returns for 1695 show James' grandfather living at 'Moncrieff of Kilforgies' land in the parish of Greyfriars. The Moncrieffs were lawyers and owned the Kilforgie estate in Perthshire. Their Edinburgh property was rented out. The tax records give us an idea of the Hutton family wealth. For tax purposes, the number of hearths (fireplaces) was used as a proxy for the size of the apartment. William is shown as having three. A dozen families lived in the same building; two lawyers and another merchant also had three hearths, as did John Clark, although the tax collector has added a cryptic note: 'hes Moe hearths but refuises to shew his hous.'[4] The other families had one or two hearths, including a widow and another woman, Jean Broun, described as being 'on charitie', so she paid nothing. This would have put the Huttons at the top of the social scale in their building, but they were by no means the most prosperous in their neighbourhood. Thomas Rigg, who owned and rented out a neighbouring building, had eight hearths in his apartment. His was clearly a superior dwelling – the smallest number of hearths of his tenants was five and 'The Lady Lenie belonging to Gosfoord' had ten.

An analysis of the status of merchants shows grandfather William heading a household of ten – his eight children, plus his wife and himself – and paying poll tax of £4 a year – a middling amount between the minimum of £2 10s (£2.50) and the highest paid by trades people of £10. Merchants were assessed on the value of their stock and William's payment indicates that his business had stock worth up to 10,000 merks* (about £500). He was wealthy enough to have subscribed £100 to the fundraising for the Company of Scotland in 1696, which mounted the ill-fated attempt to establish a Scottish colony at Darien, in Panama. His

---

* A merk was a silver coin, worth two-thirds of a Scots pound. Before the Act of Union in 1707 Scots pounds traded at 12 to 1 against sterling, the English pound.

estate at death was valued at 32,864 Scots pounds, or about £2,700 in sterling.[5]

James' father, William, was born in 1686 and followed his father into business and the public life of the commercial community. He owned a property in Gosford Close, which ran down from the Lawnmarket, the upper part of the Royal Mile just below Castlehill, to the Cowgate, the lower parallel road to the south. It was in the parish of Greyfriars, the church a little way south of the town centre, where the family probably worshipped.[6]

William was admitted to the Merchant Company of Edinburgh in 1711, signing its membership roll with an unusually elaborate signature which surely shows high self-esteem and ambition.[7] His younger brother James followed him in 1721, signing his name much more plainly. On payment of their dues – £12 Scots (£1 sterling) entry fee and £6 Scots (50p) four times a year – they received 'tickets' entitling them to call themselves 'burgess' and 'guild brother'. The Merchant Company could trace its roots to 1260 and received a Royal Charter in 1681 allowing it to incorporate and giving it considerable privileges. In return for paying extra taxes, members were granted exclusive rights to trade and import goods. The company was zealous in enforcing its monopoly, taking legal action against craftsmen who attempted to sell their wares directly, rather than through merchants, or any non-member who set up shop or tried to bring in manufactured items from abroad. In the early part of the 18th century it was particularly concerned with the number of women setting up as drapers. It permitted widows of members to join provided they carried on the business of their late spouses, but if they remarried the membership passed to their new husbands.[8]

The company was intimately involved in the governance of the city. Initially it had elected all members of the council. King James VI had widened the franchise by permitting the incorporated trades to elect some councillors, but merchants were still dominant, providing the baillies (senior councillors), lord provost (leader of the council), the treasurer and the dean of guilds, who had power over regulations and planning matters.[9] It also had charitable activities, supporting 'decayed' merchants in their old age and educating the orphans of merchants. Using a donation from the money-lender

Mary Erskine (Mrs Mary Hair) it began teaching girls, who were housed in the Merchant House, a grand mansion just off the Cowgate purchased in 1691. Later a bequest from James Watson enabled the company to start a school for boys.[10]

In addition to its civic responsibilities, the council had educational, judicial and religious power. It appointed and paid teachers in the High School and the principal and professors in the university. It elected the magistrates who presided over the burgh courts, trying petty crimes. It collected 'seat rents' from the churches and appointed and paid kirk ministers. William Hutton was elected a merchant member of the town council in 1720 and enjoyed a swift rise through the hierarchy, becoming in rapid succession a baillie, dean of guilds and treasurer. In 1723 we find him at the Canongate kirk session with George Drummond, dean of guilds, and Andrew Wardrop, deacon convenor of the council, appointing a new minister to the church.[11] His term of office as treasurer came to an end in 1724, but he remained on the council as 'old treasurer' for a year and used his position to vigorously oppose plans by the goldsmiths to extend their hall and increase the number of shops under it – doubtless at the expense of merchants.[12] The following year he became master of the Merchant Company, a position conferring considerable prestige and authority in the city. It was a natural progression from master to lord provost and had he not died he may have become leader of the council.

It is likely that he met his wife, Sarah, through his membership of the council, since her father, John Balfour, had also been a councillor. William and Sarah were married in 1720, when she was 21 and he was 33. Their first child, a boy named after his father, was born a year later, followed by another son, John, 18 months after that, but then tragedy struck the family. Their son William died at only two years old. Infant deaths were common. Sanitation was poor, epidemics were frequent and medical training had only recently been introduced into the university. There were few midwives, none of them trained. The city was three years from getting its first hospital – with just four beds, serving a population of 40,000 – there would not be a purpose-built hospital for another 15 years.

The loss may have been softened by the birth of two daughters, Isabella and Jean, soon after, but then a second blow when John died, aged just three. Sarah was already pregnant again and James Hutton was born on 3 June 1726, just six months after the death of his brother. We can imagine how much anxiety his parents must have had over this new child and how much hope they invested in him. He was the only one of his parents' male children to survive beyond infancy and named for his uncle James, who had died the year previously. Another sister, Sarah, was to arrive two years later.

In 1729, a few weeks after James' third birthday, the family were dealt a devastating misfortune: his father William died, aged just 42. No cause of death is given on the burial record, but outbreaks of diseases such as cholera, typhus and typhoid were common and spread quickly in such a crowded community. He was buried in the Canongate churchyard. His will showed that he owned a number of properties which were rented out, some of which he had inherited from his brother John, who had died the year previously. He was also owed money by men as far afield as Aberdeen and Daviot in the Highlands. It was common for merchants and others with spare cash to lend it out at interest. Often the loans were arranged by lawyers and the lenders need not necessarily have known or even met the borrowers. Altogether rents and interest on his lending brought in around £500 a year – a reasonable annual income for his widow, Sarah, and her four children to enjoy a comfortable life, but not enough to make them wealthy.[13]

Sarah must have been a robust woman. In nine years of marriage she had given birth six times, had lost two children and now her husband. His will, made some time before his death, provided that not only his wife, but her two brothers, her father-in-law and three merchant friends should act as 'tutors and curators' to the infant James Hutton. But at the time of his death William Hutton senior and Sarah's brother John were also dead. Her other brother Andrew was spending much of his time on the family estate, Braidwood at Temple, Midlothian. The burden of bringing up her children fell on her. Initially James was educated by his mother, who either engaged paid tutors to come to their home or sent the boy to one of the small private schools which operated in the city. By the time he went to the High School at the age of ten he would have been able to read and write and have a smattering of Latin and Greek. The school, then housed in a small building off the Cowgate, was a stark transition from the all-female household in which he had been brought up. Now he entered an all-male world with a reputation for rote learning, harsh discipline and violence.

The High School had a long history. Founded in the 12th century as a seminary for Holyrood Abbey, after the Reformation it was transferred to the care of the town council. It educated the nobility and enjoyed royal patronage in the reigns of Mary, Queen of Scots, and her son James VI, but when he accepted the throne of England and decamped to London many Scots aristocrats moved with him. There was a second wave of emigration following the Act of Union of 1707, which joined Scotland and England together in a United Kingdom. A decade later the rector (headmaster) was complaining: 'There are scarce any of the nobility and very few of the gentlemen of the country residing in Edinburgh and youths who attend my instruction are almost altogether the children of burgesses.'[14] The school had become middle class, largely educating the sons of merchants and the professions, although there were still a few sons of minor nobles.

The council exercised firm control, setting the curriculum, carrying out regular inspections, appointing the rector and masters and fixing their stipends and the fees they could collect from parents for extra lessons. Masters were expected not only to be competent, but to conform to the prevailing religious and political orthodoxy. The council was under the patronage of the Whig faction led in Scotland by the Duke of Argyll. In 1739 John Rae, a schoolmaster from North Berwick, was engaged to be one of the teachers, councillors having satisfied themselves that he was 'well affected to the Established Government of Church and State'.[15]

In James Hutton's time the school was divided into four classes, each led by a master who remained with the same boys throughout their four-year school career. Latin and Greek dominated the syllabus. A 'writing master' had been appointed, although attendance at his class was optional and he was not paid by the council but had to exist on the fees paid by parents. In 1715 the city treasurer was authorised to spend £20 on maps for the school, but it was not until 1742, after Hutton had left, that geography was formally added to the curriculum and two globes purchased. In 1710 'Wedderburn's Rudiments' had been replaced as the standard text for classics because 'it is a confused mass of hard Greek words in Latin characters containing nothing which is not better explained in the short compendium of rhetoric'.[16] The lawyer Henry Cockburn, who attended the school 50 years after Hutton, remembered the uniform pupils were required to wear. It probably had not changed much since Hutton's day:

> A round black hat; a shirt fastened at the neck by a black ribbon, and, except on dress days, unruffled; a cloth waistcoat, rather large, with two rows of buttons and of button-holes, so that it could be buttoned on either side, which, when one side got dirty, was convenient; a single-breasted jacket, which in due time got a tail and became a coat; brown corduroy breeches, tied at the knees by a showy knot of brown cotton tape; worsted stockings in winter, blue cotton stockings in summer and white cotton for dress; clumsy shoes made to be used on either foot, and each requiring to be used on alternate feet daily; brass or copper buckles. The coat and waistcoat were always of glaring colours, such as bright blue, grass green, and scarlet.[17]

We can imagine how uncomfortable the modest Hutton must have felt in such a conspicuous outfit.

The school regime was strict. In 1696, during the 'ill-years', a series of particularly hard winters, parents petitioned the school to delay the 7 a.m. start time during the winter months. As a concession, lessons began at 9 a.m. from November to March, but the school must have reverted to its early start because parents were again calling for later winter hours in 1723. Boys attended six days a week – although a half day was set aside for sports and play. On Sundays they were expected to attend services at the neighbouring Lady Yester's Kirk. The school was infamous for 'barrings', when boys would barricade themselves into the school to compel masters to concede a holiday. Often force had to be used to dislodge them and on one occasion a baillie (senior councillor) leading the town guard was shot dead trying to gain entrance. 'Bickers', street fights between High School boys and students from the 'Tounis College' (now the University of Edinburgh), were also common.[18]

The council set the disciplinary code and masters frequently thrashed delinquent boys. Cockburn wrote: 'Out of the whole four years of my attendance, there were probably not ten days in which I was not flogged, at least once.'

> Yet I never entered the class, nor left it, without feeling perfectly qualified, both in ability and preparation, for its whole business; which, being confined to Latin alone, and in necessarily short tasks, since every one of the boys had to rhyme over the very same words, in the very same way, was no great feat. But I was driven stupid. Oh! the bodily and mental wearisomeness of sitting six hours a-day, staring idly at a page, without motion and without thought, and trembling at the gradual approach of the merciless giant. I never got a single prize, and once sat *boobie* at the annual public examination. The beauty of no Roman word, or thought, or action, ever occurred to me; nor did I ever fancy that Latin was of any use except to torture boys.[19]

Whether Hutton enjoyed his time at the High School, or merely endured it, he does not tell us. The mastery of Latin stood him in good stead in a later phase of his education. After four years of schooling he entered the University of Edinburgh (then called the Tounis College) at the age of 14, matriculating on 29 March 1740 – a month after most other students – in the class of John Ker.[20] Edinburgh was growing in reputation as the pre-eminent science university in Britain, surpassing even Cambridge in the number and quality of scientists it educated,[21] but Hutton chose to study humanities. Ker was a classics scholar, but also taught some law. Alexander Carlyle, who had enrolled with Ker two years before Hutton, thought him 'very much master of his business', although too partial to the titled members of his class.[22]

Professors at the university, appointed by the town council and paid a small stipend, had to depend for most of their income on attracting fee-paying students. At the beginning of the academic year they held open meetings when students could hear an outline of the course to be taught and get an impression of how interesting or useful the lectures might be. If a student wanted to sign up, he (universities were all-male preserves) paid the fee – from two to five guineas, depending on the reputation of the professor – and received a 'ticket', which enabled him to attend lectures. Janitors stood at the door and refused entrance to those without a ticket. The system enabled students to pick and choose their courses. There was no absolute necessity to matriculate with the university, or to graduate. Many took a few courses and left without taking a degree.

Alexander Carlyle, although he intended to follow his father in becoming a Church of Scotland minister, also took mathematics classes taught by Colin Maclaurin. Playfair suggests that Hutton, too, attended some of Maclaurin's lectures. Besides being an eminent mathematician in his own right, Maclaurin was a disciple and protégé of Isaac Newton, whose scientific method and revolutionary ideas he promoted. His students considered him to be a gentle man and a brilliant teacher, and his seminars attracted people from the town as well as students and were said to be 'amusing and instructive to a promiscuous audience of both sexes. His style of lecturing is represented to have been uncommonly interesting, which was in shining contrast to the system that prevailed among most professors of dictating their lectures to their classes.'[23]

In 1743, Hutton signed up for the course in logic and metaphysics taught by John Stevenson, who was also challenging old ideas and introducing new thinking, particularly the writing of the philosopher and political theorist John Locke. According to one researcher of the period: 'The best English authors were little known in Scotland. Men who in life spoke the broadest vernacular could not easily read or write in English, which to them was a foreign tongue in which they might make more blunders than in school-learned Latin. Stevenson was among the first to point out the works that ought to be read and studied in order to improve the taste; or to specify authors whose writings were considered as models after which a young writer should copy.'[24] Hutton's later interests suggest that Stevenson helped to open the young man's mind to a world much wider and more exciting than he had learned about at school.

Playfair, however, points to a different effect. Stevenson had apparently used as an illustration of some problem of logic the fact that gold is resistant to attack by nitric acid or hydrochloric acid, but when they are combined into aqua regia it can be dissolved. Hutton 'became from that moment attached to [the study of chemistry] by a force that could never afterwards be overcome. He made an immediate search for books that might give him some farther [sic] instruction.'[25] Hutton left the university after three years, apparently without taking a degree – his name does not appear in the Laureation (graduation) album[26] – although this was not uncommon.

Chapter 3

He shall not commit the filthy crimes . . .

BIOGRAPHERS OF JAMES Hutton have written-off his first period at university as a youthful misstep and, indeed, it was followed by a dead-end job. It was a natural progression from the High School to the college, a few minutes' walk away in his hometown. Many of his contemporaries made the same move and went on to lead useful if unspectacular lives as merchants, lawyers, doctors, church ministers or soldiers. But Hutton had been inspired. He had spent three years in one of the leading educational institutions of Europe and was exposed by men like Maclaurin and Stevenson not only to new ideas but to new methods of inquiry. Playfair tells us that Hutton was never a mathematician – and the matriculation records do not record him signing up for Maclaurin's three-year mathematics course – but he was influenced by Newton's experimental method and relentless curiosity, which Maclaurin's animated delivery brought to life. Besides mathematics, the professor taught experimental philosophy, surveying, fortification, geography, theory of gunnery, astronomy, and optics, so it is possible that Hutton attended one of these courses.[1] He was also influenced by Stevenson's teaching. Hutton's papers on philosophy, published decades later, show the strong influence of Locke's 'empiricism' – the belief that knowledge starts with experience.[2]

However, as important as the opening of his mind were the friends he made. It started on the first day, when matriculating alongside him was John Davie. Little is known of Davie's early life beyond that he was six months younger than Hutton and his parents, Adam Davie and Eupham Fyfe, had him christened in Canongate Kirk, in the east of Edinburgh Old Town. Given his age and the location, it is possible that he also attended the High School.

The two quickly became friends and Davie, despite signing for Professor Ker's class, also developed an interest in chemistry. Other contemporaries included John and William Hall, two sons of Sir James Hall of Dunglass, who was a farmer and coal owner with an estate in Oldhamstocks, East Lothian. Sir James' father was one of several Scottish landowners who had bought Canadian baronetcies from the government in 1687. Since Nova Scotia was at that time under French control, it was a title with no meaningful geographic connotation. John Hall succeeded to the title while he was at university. Another of Hutton's fellow students was James Pringle of Stichill, a cousin of the Hall brothers.

As the son and grandson of burgesses Hutton could have followed his father into membership of the Merchant Company, but either he chose not to or did not have the opportunity. His mother, Sarah, was never admitted to the Merchant Company, indicating that she did not carry on the family business. She may have been forced to sell it.[3] Instead, after leaving university, the 17-year-old James either opted for, or was pushed into, choosing a legal career. He became an apprentice in the office of George Chalmers, a Writer to the Signet (solicitor), who was a relative of his mother.[4] It was the first rung on a ladder to a profession in the law which would have given him a certain social status and a comfortable income. Apprentices had to get the permission of their fathers to enter into the binding contract and James is listed as 'son of William', even though his father had been dead for 14 years. Another of his student friends at university, John Bell, who was later to become Hutton's lawyer and man of business, had also entered into a legal apprenticeship, although with a different master.

Apprentices spent long hours copying out neat drafts of legal documents, whose language must have appeared archaic even to an 18th-century youth. Sitting at a desk all day was a backward step – too much like school. Playfair, in a rather picturesque image, shows us Hutton neglecting his copying duties to amuse his fellow apprentices with chemistry experiments, although it seems unlikely that he would have been able to smuggle chemicals and apparatus into a lawyer's office unnoticed. It is more likely that this happened in Hutton's next apprenticeship and that Playfair had misremembered the anecdote. Hutton lasted two years of his three-year legal indentures before it became obvious that he was not destined to be a lawyer. In Playfair's words, Chalmers 'with much good sense and kindness, therefore, advised him to think of some employment better suited to his turn of mind, and released him from the obligations which he had come under as his apprentice'.[5] James left the firm by mutual consent.

His next move was surprising: he went back to being a student, this time of medicine, and signed a five-year apprenticeship agreement with a doctor. Playfair tells us that it was because the study of medicine was 'most nearly allied to chemistry' and it is true that Andrew Plummer, professor of chemistry, was in the medical faculty at the University of Edinburgh. Whether Hutton would have found his lectures interesting and useful is doubtful. They were said to be more 'pharmaceutical than theoretical, and notoriously uninspired'.[6] Hutton also studied anatomy, attending the courses of Alexander Monro*, founder of the medical school, signing up for his classes in 1745 and 1746.[7] Monro taught anatomy by allowing his students to watch him dissecting a body, but shortage of corpses was a constant problem and the medical colleges had to warn their students and apprentices against grave robbing. His lectures, delivered in Scots rather than Latin, which had been the tradition, attracted large numbers of paying attendees.

Monro had been heavily influenced by his experience of medical education in Europe. After attending lectures in London, he had studied in Paris at Le Jardin du Roi, a 'physic' garden which also hosted anatomy and chemistry classes in its buildings. At the Hôtel-Dieu, the oldest and one of the largest of the Paris hospitals, he 'walked the wards' observing clinical practice and operations. This combination of theoretical teaching and practical medicine was continued at Leiden, in Holland, where he studied under Hermann Boerhaave, whose revolutionary methods, combing the practical and the theoretical, had made the university a model for enlightened education. Under Boerhaave students of all nationalities and religions were welcomed and given 'great liberty, the freedom of

---

* Now known as *Primus,* to distinguish him from his surgeon son of the same name, called *Secundus,* and his surgeon grandson of the same name, called *Tertius.*

thinking, speaking and believing. Unlike other medical teachers, he stimulated his students to think, observe and experiment themselves, rather than apply ready-made and generally accepted courses of action.' Unlike traditional university teachers, he never reproached his students for disagreeing with him and was prepared to learn from them.[8] Monro and other professors who had studied at Leiden brought this openness to Edinburgh.

In parallel with his university course, Hutton worked under Dr George Young. In contrast to Monro, Young had learned his trade by being an apprentice rather than by academic study, although he later received a degree in medicine from the University of St Andrews *in absentia* in recognition of his published work on the treatment of dysentery and his professional standing as a doctor. By the 1740s he was well-established and prosperous, with a spacious apartment in the Lawnmarket.[9] He had a medical practice and an apothecary's shop (which necessitated getting his 'burgess ticket') and was admitted to the Royal College of Physicians and the College of Surgeons in Edinburgh. He was a thinker and researcher as well as a practical doctor, having given a series of lectures emphasising the importance of observation and experience alongside theoretical knowledge. His treatise on opium challenged accepted medical opinion, acknowledging its therapeutic uses, but drawing attention to its risks. Young had been a member of the Rankenian Club, a group of young intellectuals which also included university professors Colin Maclaurin and John Stevenson, and Robert Wallace, a mathematician and minister at Greyfriars church. David Hume had been a member while a student at Edinburgh University.[10]

Hutton signed his apprenticeship deed in 1745, binding himself for five years and paying a fee of £50 a year.[11] Apprenticeship indentures were legal agreements between masters and apprentices, guaranteeing their good behaviour. One from the archives of the Royal College of Surgeons of Edinburgh demands:

he shall not reveal his Master's Secrets in his Arts, nor the secret Diseases of his Patients, to any person whatsoever, nor shall have any Patients of his own under Cure, upon any pretext whatsoever [but also] he shall not commit the filthy

Crimes of Fornication or Adultery, nor play at any Games whatsoever; And that he shall not be drunk, nor Night-walker, nor a haunter of Debaucht, or idle Company, nor go to Ale-houses, nor Taverns, to tiple or drink with any Company, whatsoever.[12]

Dr Young had two sons, both of whom he intended should follow him into medicine. Thomas, the younger of the two was the same age as Hutton and did indeed become a celebrated doctor and professor of midwifery. George Young, eight years older than his brother, had served his apprenticeship in his father's shop and was then sent to Paris to 'walk the wards' – gain practical experience in the hospitals. After a year he transferred to Leiden, supposedly to continue his studies, but was in fact spending his allowance on wine, spirits, other luxuries and on courting a local woman, whom he took to Colchester, Essex, to be married, claiming that he had his father's permission. The following year a child was born, but by this time George Young junior had been summoned home by his father, who started legal actions to declare the marriage invalid and was in turn sued by his bankers, whom he refused to reimburse for the money they had advanced his son in Leiden.[13] The case caused a sensation and James Hutton would have been well aware of the details, since it was still going through the courts when he was in Dr Young's service.

Despite the prohibitions in apprenticeship contracts, medical students were notorious for illicit relationships. 'Sex was available in brothels, frequently mentioned in student diaries, although not in letters to their fathers,' writes one historian.[14] Silas Neville, who was an Edinburgh medical student in 1771, described visiting three brothels – 'not from any bad intention, but merely from curiosity,' he added, by way of mitigation. 'In an Edinburgh brothel dirtiness and vice are combined. The last indeed we visited is an exception; the rooms were tolerably clean. It is kept by a Mrs Jap, an exact old Bawd. May the sight of vice always excite abhorrence in me!'[15] Elsewhere in his diary Neville makes clear that he was on occasion able to overcome his disgust, although with a clear conscience: 'every connexion I have had with such [common] women was accompanied with my earnest endeavours for their reformation'.[16]

Whether Hutton had his first experience with the opposite sex while he was a medical student, we cannot say. Unlike students from outside the city, he was probably still living with his mother and sisters, which might have inhibited his activities. The fact that his three sisters all remained unmarried suggests that the family lived a quiet life – there were no house parties where the girls might meet eligible bachelors. But through another of his university friends James, now 19, was introduced to a more congenial and stimulating social world.

Hutton may not have met John Clerk before they both took Monro's anatomy classes in 1745 and 1746.[17] Two years younger than Hutton, Clerk had been brought up in Midlothian and had gone to school in Dalkeith. His father, Sir John Clerk of Penicuik, wanted him to become a doctor and wrote him four pages of advice exhorting him 'to walk and behave in all things as in the presence of Almighty God . . . Keep yourself from bad habits and bad company for your manners will be easily corrupted . . . Cultivate an inviolable honesty in all your dealings, for a man who is once caught in dishonesty will have the fate of a liar, never again to be either trusted or believed.' As far as his studies were concerned, he should 'neglect no opportunity of studying anatomy with the best masters and in your private practice beware of the barbarity of raising dead bodies. The burial places of the dead were in all ages especially sacred and on no account ought to be violated.' He should keep up his mathematics and take a course in experimental philosophy. Finally, Sir John advised: 'I own that a man may be a great physician without the knowledge of the Greek language, but you will never have the honour of being reputed a scholar without it.'[18] Hutton, of course, had no father to guide him, but Sir John's ambitions for his son were of no avail – he abandoned his medical career soon after leaving university.

The Clerks of Penicuik, a very large, distinguished and erudite family, were to play an important role in Hutton's life. Head of the family was Sir John, lawyer, antiquary and sometime politician, known to his close friends as 'the baron' because of his role as one of the barons of the exchequer. He had been married twice and had nine sons and seven daughters. He inherited extensive estates, which he improved by planting trees and modernising the tenancies, and owned coal mines. He was a cultured man, interested in archaeology, poetry, music, and literature, as well as politics and economics. He had been educated at Leiden and undertaken a Grand Tour of Europe, which included studying in Rome – music under Arcangelo Corelli and painting with Carlo Maratta. He was a fellow of the Royal Society in London and became president of the Edinburgh Philosophical Society. There were regular house parties in Penicuik House and at a second home Sir John had built at Mavisbank, Loanhead.[19] Both were within an easy ride from Edinburgh, and it is clear from a later letter of Hutton's to George Clerk, John's brother, that he had been a guest at Penicuik and elsewhere and had become friends with 'the baron', Lady Clerk and with several of John Clerk's brothers and sisters. He had found their company and their conversation stimulating, entertaining and exciting – everything that his home environment was not.[20]

But life was not all parties for Hutton at this time. Edinburgh was in turmoil in 1745. In July Bonnie Prince Charlie, the 'Young Pretender', had landed on the north-west coast of Scotland to lead an insurrection aimed at putting his father on the thrones of Scotland and England. The country had been convulsed by Jacobite rebellions twice before, in 1689 and 1715. On this occasion Britain was fighting a Continental war and most of its regular troops were abroad. By September the prince had raised an army and was threatening the Scottish capital, which was only lightly defended. Colin Maclaurin, Hutton's one-time mathematics teacher, was put to work shoring up the city's defensive walls and the lord provost, George Drummond, formed a citizens' militia, which included some of Hutton's former university colleagues. Whether he marched with them as they left the city to meet the rebels he does not tell us, but in the event there was no fight; Edinburgh fell to the Jacobites without a shot being fired. Many of the better-off merchants and professionals left the city for the relative safety of their country houses or estates. The Hutton family owned two farms in Berwickshire. They may have moved there, but we have no evidence.

Ten days later the rebels defeated a British army at the Battle of Prestonpans, a short way outside the city, causing consternation in Edinburgh. The Jacobites held the city, but they could not take the castle, and the two banks, Bank of Scotland and the Royal Bank, had hastily transferred their gold, silver and notes there for safe-keeping. The prince had virtual control of Scotland, but his ambition was to take London and he began preparing for a march into England. To raise cash a tax was imposed on the capital of half-a-crown in the £ (one-eighth) of all rents, which all householders and businesses had to pay. Additional demands were made on large landowners. Sir John Clerk, who with his wife had fled Penicuik House, was forced to provide 6,000 stone of hay (about 38 metric tonnes) and 75 bolls of oats (about 16 cubic metres) 'on pain of military execution, which was understood to be the quartering of some Savage Highlanders upon us'. Despite providing the food and fodder at a cost of £200, 'they quartered themselves frequently upon us and our tenants, so that the family I had left at Penicuik was obliged to entertain some of their chiefs – three several times and frequently 16 or 20 at a time'.[21]

In November the Jacobites left Scotland to march south, taking Carlisle, before halting near Derby and then retreating. There was fear in the capital that they would return to Edinburgh, but they passed west of the city, defeating another government army at Falkirk and occupying Inverness before their final defeat at Culloden in April 1746.

Chapter 4

All that . . . is absurd and false

PLAYFAIR TELLS US that Hutton was with Young for only three years, but his apprenticeship contract was for five.[1] After completing Monro's anatomy course he left Edinburgh at the end of 1747 for Paris.[2] As with other periods of Hutton's life we have very little hard evidence. Playfair's is the only written source for Hutton having been in France at all and searches – for this book and by other researchers before me – have not discovered any confirmation. The rest of this chapter is therefore based on conjecture; what was happening in the city at the time and the recorded experience of others in a similar situation. Yet Paris appears to have been formative for Hutton, exposing him to new experiences and ideas and possibly introducing him to people who may have had an influence on his later theories.

Britain was still at war with France over the leadership of the Habsburg empire and a peace treaty was not signed until the following year, but travellers were still able to visit the country. By the middle of the 18th century, Paris was the pre-eminent place for the study of medicine, offering unrivalled opportunities to get both theoretical teaching and practical experience. In contrast to Edinburgh, which had a population of less than 50,000 and only one hospital, Paris was home to 600,000 people and had 23 hospitals run by the Catholic Church, central and local government. In addition to the huge Hôtel-Dieu and Notre Dame, which could house 2,500 sick and poor people (rising to 4,000 during epidemics), there were specialist smaller institutions. L'Hôpital de la Charité, run by the monks of St Jean de Dieu, was cleaner, admitting only one patient for each of its 200 beds, Les Invalides catered for army veterans and Pitié-Salpêtrière for women.[3]

The University of Paris, one of the oldest in Europe, was jealous of its reputation as the centre of medical education, but its conservatism had been challenged by progressive scientists and doctors, who had followed other European centres of learning to establish a Jardin des Plantes – a botanical garden to cultivate medicinal herbs. Despite opposition from the university, they obtained royal patronage – signalled by a change of name to Le Jardin du Roi – and were able to appoint professors and demonstrators (assistant professors) of botany, chemistry and anatomy. To the fury of the university they staged popular lectures, which were free of charge, open to all and in French, rather than Latin. In 1739 Georges-Louis Leclerc, Comte de Buffon, was appointed as director of Le Jardin. Still only 32 he had established a reputation as a mathematician, natural scientist and free thinker. He was a friend of the philosopher Voltaire and considered himself a deist – he believed in God but did not follow the biblical narrative of the Creation. Natural processes, he believed, had to be understood through evidence, not from blindly following scripture.

Buffon set about enlarging the garden by buying adjoining properties and building a large amphitheatre and natural history museum. He also made the garden a centre of research by recruiting leading scientists from all over France and abroad. They promoted modern medical theories, such as the circulation of the blood, which were shunned by the university.[4] The English doctor William Harvey had described the function of the heart, the arterial and vascular systems in 1628, but many institutions, including the University of Paris, still taught the theory of the ancient Greek physician Galen, that blood passed directly from one side of the heart to the other. During Hutton's time the professor of anatomy was Jacques-Bénigne Winslow, originally Danish, but now with French nationality. He had published a treatise on the circulation of blood in foetuses and a three-volume textbook of anatomy which described the circulatory system in detail. It had been translated into English and republished in London.

By the time Hutton arrived, Le Jardin du Roi was educating dozens of French and foreign medical students, although to pacify the university it was not allowed to award degrees – students had to go elsewhere to graduate. Hutton probably attended lectures there, as Monro had done, although, since the institution kept no records of its students, we do not know for certain. He could have continued his surgical and medical training by attending ward rounds and watching operations in hospitals. 'Walking the wards' was an established way of gaining first-hand experience, but it involved an early start. Students had to get from their lodgings to the hospital before ward rounds began, usually before 9 a.m., and some physicians liked to start at 6.30 or 7 a.m.[5]

The Scottish surgeon James Houstoun gained access to the Hôtel-Dieu around 1714 after studying in Edinburgh for two years and Leiden for three. He later wrote: 'Paris is certainly the best place for learning the practical part of anatomy and surgery, from the frequent opportunities of seeing chirurgical operations of all sorts perform'd in the hospitals.'[6] Some students found lodgings with the surgeons who were supervising their medical studies. Sauveur François Morand, chief surgeon at the Charité hospital, took in more than 70 foreign students *en pension* over a 20-year period up to 1746. Among the 'many from England and Scotland' was the military physician John Pringle, uncle of Hutton's university contemporary James Pringle.

Hutton's time in Paris was his first journey outside Scotland and, for a young mind open to new experiences, must have been life changing. It would also have proved challenging. We don't know whether Hutton had learned any French before he made the journey; the High School appears to have taught only Greek and Latin. Playfair's assertion that Hutton spent two years in Paris would have made him unusual. Courses at Le Jardin ran from November to June. Most students spent less time in the French capital – George Young, son of Hutton's Master, was there for a year and Alexander Monro less than that. It is possible that Hutton spent some time learning French before embarking on his course.

We cannot be sure about any of the details of Hutton's stay. None of his surviving letters or publications mention it and formal matriculation was not required for Le Jardin du Roi, the university or the hospitals. Many lectures were free and open to the public, although private surgeons and physicians also gave courses for paying students. We do not know where he lived. The police, based in the bastion of Le Châtelet, apparently kept records of foreigners living in the city, but these were destroyed during the French Revolution or the siege of the city in 1870. Le Jardin was in the Quartier Saint Germain, on the left bank of the River Seine. This was only a short walk from the Île de la Cité, where some of the main hospitals were situated. The area was fashionable, prosperous and had a bustling night life with theatres and cafés. Edinburgh must have seemed grey and dull by comparison.

Contemporary accounts of the cost of living in Paris vary widely. Some medical students spent lavishly, enjoying the social life and visiting the tourist sites. Others lived more frugally on as little as £10 a year. Hutton's mentor, Dr Young, had allowed his son George £100 a year when he had studied in Paris nine years before, writing to his Paris banker: 'he has been used to no degree of luxury or elegance', that his usual fare was 'pottage and milk' and that he was to be 'kept free of company until his studies are over'.[7] That banker was Aeneas MacDonald, who acted as Paris agent for many Scottish banks, but his services were not available to Hutton. MacDonald was a Jacobite and one of the Seven Men of Moidart, who travelled from France to Scotland with Bonnie Prince Charlie in 1745. He was captured after Culloden and by the time Hutton was in Paris, he was on trial in London. Hutton may have benefited from Young's other contacts in finding lodgings and a surgeon to introduce him to the hospitals.

At Le Jardin, the large auditorium, which could seat 600, was used for anatomy demonstrations in the winter and botany and chemistry classes in the spring and summer, when the temperature made dissecting corpses impractical. Hutton most probably attended the chemistry lectures of the brilliant and eccentric Guillaume-François Rouelle. Until Rouelle arrived at Le Jardin in 1742, chemistry had been taught as an adjunct to pharmacy and medicine, mainly by men who had trained as apothecaries and who emphasised the properties of herbal drugs and clinical preparations. Rouelle, by his colourful personality and explosive temperament, revolutionised the teaching of chemistry, attracting big audiences to his free public lectures and earning himself a national reputation.

He was not afraid to challenge orthodoxy, even his own superior, the elderly and unenthusiastic Louis-Claude Bourdelin, reputedly beginning his lectures: 'Gentlemen, all that *monsieur le professeur* has just told you is absurd and false, as I will prove to you.' His lectures were often flamboyant, and usually illustrated with experiments. One of his students described listening to him: 'when Rouelle spoke, he inspired, he overwhelmed; he made me love an art about which I had not the least notion; Rouelle enlightened me, converted me; it is he who made me a supporter of that science [of chemistry] which should regenerate all the arts, one after the other . . . without Rouelle, I would not have known how to look above the mortar of the apothecary.'[8] The lectures would have been an effort for Hutton to understand, even if he was fluent in French, because Rouelle spoke with a thick Normandy accent.

Rouelle departed from the earlier syllabus to include general chemistry, the classification of acids, bases and salts and the phlogiston theory (see Glossary of scientific terms), which held that all flammable materials contained 'phlogiston', a substance without colour, odour, taste, or weight that was given off in burning. It would be another half-century before another French chemist, Antoine Lavoisier (who had also been a student of Rouelle), disproved it by experimentation. Rouelle became one of the foremost teachers of science in France, educating a generation of chemists and non-scientists including the philosophers Rousseau and Diderot. According to Diderot, his lessons attracted 'a quarter of the city' and all classes of society, without forgetting 'the children of the nobles who wanted to learn'. Even with its large capacity, the lecture theatre was sometimes too small for the crowds who came to listen.[9]

Rouelle left no transcripts of his lectures, but from the notes taken by a student we can piece together the ground he covered. His course at the Jardin du Roi included about sixty lessons and took place over a three-year cycle, describing the 'three kingdoms of nature' – animal, vegetable and mineral.[10] His first lesson was devoted to the definition of chemistry and its role in the explanation of major physical phenomena such as volcanoes, lightning and earthquakes. He also dealt with the practical applications of chemistry in cooking, painting, dyeing, varnishing, the manufacture of glassware and metals. He described the four elements: phlogiston or fire, earth, water and air and suspected that there may be a fifth.

Rouelle was above all an experimental scientist and the rest of the course reported in the student notes is accounts of many experiments: fifty-six for the vegetable kingdom, eleven for the animal kingdom and one hundred and fifty-nine for the mineral. Sometimes he performed an experiment in front of his students, or if this was not practical, exhibited the results.[11] If Hutton did indeed take this course we can see how Rouelle's practical mind might have influenced his approach to science. Aspiring doctors probably only took part of Rouelle's very comprehensive programme, but Hutton's interest may have been stimulated enough for him to stay longer. It could have been his first introduction to geology, in which Rouelle covered bitumen, sulphur, succinic and mineral acids. Under the heading stones and earths, he described limestones, gypsum, clays, quartz, 'spath' (fluorspar) and 'apyres' (feldspath). He also dealt with metals and 'half-metals' and alchemy.

Attempts had been made to discredit the philosophical system of alchemy and particularly the search for the Philosopher's Stone, a mystical element which was supposed to facilitate 'transmutation' – turning base metals into gold. But some serious scientists, including Newton, still studied and taught it. Rouelle appears to have conducted experiments in a secret laboratory. He is reported to have told his students that the Philosopher's Stone was nothing more than the result of a fermentation of gold with mercury – not common mercury, but a special mercury oversaturated with phlogiston. However, he discouraged his students from conducting such experiments themselves because success in so expensive an undertaking was very uncertain 'without a sure guide to lead one through an operation that is preserved only in oral transmission'.[12]

The introduction of Rouelle's course on the mineral world included a long essay about geology and the history of the earth, claiming that metals and semi-metals like antimony, bismuth, zinc, and cobalt were not mixed at random in the earth, but were organised in a very symmetric manner. In order to justify this assumption he suggested there were two general strata. The *terre ancienne* was the primitive earth that had always existed; the *terre nouvelle* was a layer laid down by later geological factors. Although the *terre ancienne* had also been altered by volcanic activity, the crystalline structure of rocks had remained unchanged, whereas in the *terre nouvelle* minerals were dissolved and transported in saline form to the sea. A curious aspect of Rouelle's theory was his assumption that a large body of water was present in the centre of the Earth, which activated most of the transformations; together with a central fire it caused all the changes that took place in the interior of the globe.[13]

If Hutton heard this lecture, it would have been his first exposure to the grand theorising which was becoming increasingly common among Enlightenment thinkers as they sought to explain the development of the Earth. Rouelle prided himself on being a practical scientist: 'La chimie ne cherche pas de vains raisonnemens, elle cherche des faits' ('chemistry does not look for vain arguments, it seeks facts'). But in the study of the earth some 'facts' could not be observed, either because the events they described had happened long ago, or in a place (such as the centre of the globe) to which we had no access.

Rouelle, who also gave private lessons at his laboratory in the Rue Jacob, was the latest in a line of charismatic teachers to expand Hutton's mind. His motto *'Nihil est in intellects, quod non prius fuerit in sens'* (literally, 'there is nothing in understanding, which is not first sensed'), reinforced the message that Hutton had received from his lecturers at Edinburgh, that knowledge had to be gained from experience, observation and experiment, not handed down by unchallengeable authorities. Paris also exposed Hutton to other influences. Although he was director of Le Jardin, Buffon did no teaching, but his ethos of learning from nature pervaded the organisation and Hutton may have been aware of his deism and his interest in the origin of the world.

Buffon's *Theory of the Earth* – first in a long series of books on the natural world which occupied him for the rest of his life – did not appear until the year Hutton left Paris, but it caused a sensation in France and abroad, so it is almost certain that Hutton was aware of it. Buffon proposed that the planets had been formed by the debris from a collision of a comet with the sun. Over a long period, the Earth had cooled to the point where it could support life. From experiments with metal spheres, Buffon calculated that this process could have taken over 70,000 years. This may strike us as an impossibly short period, but at the time, the accepted age of the Earth, derived from textual analysis of the Bible, was less than 7,000 years.

The book, the first of Buffon's many contributions to the debate on the origin of the Earth, marked him out as a revolutionary.

Wandering in the garden would have provided Hutton with respite from his studies and the hurly-burly of the busy city. It would also have shown him how horticulture could serve science and commerce. Edinburgh had a small physic garden, founded in 1670 by the doctors Robert Sibbald and Andrew Balfour, both of whom had studied in Paris. While working in Dr Young's apothecary's shop, Hutton may have had occasion to visit it to obtain herbs. But Le Jardin du Roi was on a much larger scale and cultivated a far greater range of plants. Buffon was anxious that the garden should serve a practical as well as an aesthetic purpose. This was a lesson not lost on Hutton in later life.

It would also be surprising if the gregarious Hutton did not make friends with some of his fellow students, as he had done in Edinburgh. Paris was attracting bright young men who wanted to obtain a scientific education. Among these was Nicolas Desmarest, who was less than a year older than Hutton and like him hungry for new knowledge. Born to a poor family in Soulaines, a small town in the Aube, east of Paris, he had started his education late, but encouraged by his brilliance and his enthusiasm, his teachers in the college in Troyes had waived their fees to teach him and then encouraged him to go to Paris to further his interest in physics and mathematics. There he had supported himself by tutoring while he attended classes and broadened his interest in science. He came to wider attention in 1752 when he won an essay competition with a paper arguing that Britain and France had once been connected by a land bridge.[14] Rouelle's popular and free lectures at Le Jardin would have appealed to the young Frenchman as much as they would to Hutton. We have no proof that they ever met or kept in touch after leaving Paris, but an incident 40 years later suggests that they might have done (see Ch. 12).

Hutton might have stayed in Paris for a second year, or since he was still apprenticed to Dr Young, might have returned to Edinburgh to continue his practical education and work on his thesis. What we do know for certain is that in the summer of 1749 he travelled to Leiden, had his thesis printed by Wilhelm Boot[15] and on 14 August he enrolled at the university.[16] A month later he presented the thesis

'*Dissertatio physico-medica inauguralis de sanguine et circulatione microcosmi: quam, annuente Deo optimo maximo*' ('*Dissertation on the Blood and the Circulation of the Microcosm*'). It was dedicated to two of his medical mentors, Dr George Young and Dr John Stevenson.[17]

Written in Latin, the 10,000-word paper is the first demonstration Hutton gave of how his thinking and his approach to science had been influenced by the teaching he had received. It is not merely a medical description of the circulation of the blood. Hutton had been trained to think of knowledge as the product of observation, experiment and experience and he had been taught by Edinburgh professors Alexander Monro, Andrew Plummer and Andrew St Clair,[18] all of whom had studied under Boerhaave at Leiden. Although he had died in 1738, Boerhaave's influence was still strong in the university and Hutton's thesis had to appeal to his Leiden examiners, so he followed Boerhaave's models for the working of the body[19] and showed his knowledge of the literature by citing Boerhaave's book *Elements of Chemistry* and his published medical lectures.

Hutton's choice of subject was possibly inspired by the anatomy course he had attended in Paris and Winslow's textbook. But he also seemed to have been taken by the idea of a self-renewing system.

> The blood is continually being used up, not only in the performance of its tasks but also by its very nature and constitution and would be, if left to its own inevitable fate, swiftly destroyed, it must then be re-supplied constantly by matter suitable for making up for its losses.[20]

Hutton then links the internal circulation of the body to the external circulatory systems of the vegetable and animal worlds which supply man's food. This 'matter' – the 'raw juice of the earth' – cannot by itself produce blood, but must be processed by the body. 'Thus are the very nice circulations of the macrocosm accomplished.' This concept of a perpetual system, self-renewing in the service of mankind was to find later echoes in his work.

His use of the word 'microcosm' as a synonym for the human body asserts his belief in 'order and systematicity, cosmos as opposed to chaos'.[21] The analogy of the 'microcosm' of the human system with the 'macrocosm' of the Earth system (or even of the planets) was not new. The Roman philosopher, poet and teacher Seneca, whom Hutton had surely studied during his Latin classes at the High School, had likened the circulation of blood to the flow of rivers. The idea was widely adopted during the Renaissance. It was an analogy which Hutton was to use again in his later writings.[22]

The inclusion of Almighty God in the title ('*Deo optimo maximo*') suggests that although he had been heavily influenced by modern ideas of systems and processes, some of them repetitive and seemingly eternal – and perhaps the deism of Buffon – Hutton was still a believer in a divine presence behind natural phenomena. This might just have been presentational. Leiden, for all its free-thinking and encouragement of religious debate, was still a Protestant university. It would have been a prudent precaution to forestall accusations of atheism – a common and often effective way of silencing those whose temporal explanations of natural processes contradicted those of the Church – but other evidence suggests that Hutton was a believer, despite his move towards rationality. This would not have been unusual among scientists. Colin Maclaurin, a son of a kirk minister who had himself considered a profession in the Church rather than as a university teacher, firmly believed in God[23] as did Newton.[24]

Hutton begins his dissertation in self-deprecating tone:

> I do not know to what extent this theme may exceed my abilities; and, indeed, working with such meagre resources of time and personal talent, I would not have dared openly to present these disordered pages; however, necessity and the customary procedure in similar circumstances have driven me to it, although reluctantly: and let no one find fault with me for having made an effort in an important matter; but, rather, let this effort, which is almost extemporaneous and of little significance, be received with a measure of good will.

But his thesis is anything but timid: he challenged orthodoxy and took sides in a current medical controversy.

He starts by describing the red and white corpuscles of blood as seen under the microscope. This was not new; the corpuscles had been discovered by Jan Swammerdam, himself a Leiden graduate, half a century earlier, but Hutton demonstrates that he has observed this for himself, not merely read it in a textbook. Similarly, he goes on to describe different characteristics of blood when clotted, when diluted, when combined with reagents such as acids and alkalis. He does not say specifically that he has carried out these experiments himself, but it is clear from the context that he has either done so or observed them. Did Hutton have access to a laboratory in Paris? (Or did he return home to conduct the experiences in Dr Young's apothecary's shop?) Or did Rouelle perform the experiments for his students in the private classes he gave in his laboratory?

The paper contains a lot of chemistry and, interestingly, in a throwback to Stevenson's analogy of two acids combining to dissolve gold to show that a compound of elements might behave differently from its constituent parts (see Ch. 2), Hutton develops this idea: 'Although every pure principle enjoys its own particular powers, still, when it is united with other things in a compound of blood, it cannot act independently upon external bodies applied to it but only through the combined capacity of their mutual union.'[25]

Hutton appears to accept the assertion by Boerhaave and Rouelle that all chemical reactions could be explained by the interaction of the four 'elements' – fire, air, water and earth. Earth in this context is not soil nor a synonym for the planet, but a theoretical constituent of all matter – a concept inherited from the Greek philosophers. Hutton is not satisfied merely to accept its existence and challenges the timidity of existing science to define it:

Thus far the sterile and obscure nature of the subject-matter has either scared off the chemists or induced them to neglect it in favour of some topic with more attractive characteristics; and this neglect perpetuates great ignorance of chemical mixture of bodies, and a great obstacle to our understanding of this earth is posed by the uncertain notion we have of it, the definition consisting, as it were, of two merely negative and passive items.

There are digressions into discussion of the nature of earth and water before he goes on to talk about the mechanics of the circulatory system, the dilution of air in the blood. He then discusses at length illnesses which might be caused by 'bad conditions of the blood' and their possible alleviation. Again, Hutton does not specifically say he has witnessed these medical conditions or their treatment, but the context implies that he has (by 'walking the wards'?). He demonstrates his knowledge of anatomy with a discussion of the endocrine and respiratory systems and the functions of the kidneys and liver.

Hutton is 23 by this time, still young, although probably older than most candidates presenting for a medical degree, but he does not shy away from controversy. He goes against Boerhaave in his use of both fire (heat) and solvents in his analysis of blood, which the great teacher believed would have so changed the subject under examination as to render the results invalid. And he sides with Frederik Winter, a Leiden professor who may have been one of his examiners, in explaining the action of the muscles. Winter was involved in a dispute with his fellow professor, Wouter van Doeveren, over whether the muscles responded to chemical or some other stimuli. Hutton believed that the stimulus was 'some active and potent cause which is, perhaps, entirely different from any chemical, mechanical law thus far known to us'. Donovan and Prentiss, modern researchers who have analysed the dissertation, argue that Hutton's allusion to recent work on static electrical charges, although vague, showed that he was aware of contemporary research. They conclude:

> It is this confidence in the theoretical possibilities of chemistry that distinguishes Hutton's dissertation and sets it apart, if only in conception, from the works of the authors he read and the teachers he studied under in Edinburgh.[26]

Hutton was successful and was awarded the degree of Doctor of Medicine.

Chapter 5

More keen of martial toil . . .

THEN CAME ONE of the most mysterious and most important periods of Hutton's life. After graduation he returned not to Edinburgh, but to London, where he stayed for six months or longer.[1] Why London? It could have been an entirely chance destination. In the days before regular ferry services travellers took the first available ship with room for a paying passenger, so it may have been that there was no imminent sailing from Leiden to Scotland. Or it could have been a deliberate choice. If it was, why did Hutton need to go there?

Playfair offers an unconvincing explanation, suggesting that the prospects for a newly qualified doctor establishing a viable practice in Edinburgh were not promising. 'The business there was in the hands of a few eminent practitioners who had been long established; so that no opening was left for a young man whose merit was yet unknown, who had no powerful connections to assist him on his first outset, and very little of that patient and circumspect activity by which a man pushes himself forward in the world.'[2] An aspiring physician needing to attract paying patients would have had more opportunity in London than in the much smaller Edinburgh, but Hutton had many more contacts to help him in Edinburgh than he did in London. There is no evidence that Hutton tried to embark on a medical career at all. Indeed, Playfair had already hinted as much when he suggested that his principal reason for turning to medicine was to study chemistry, rather than to become a doctor.

Whatever took Hutton to London, the experience unsettled him. He does not seem to have known what he wanted to do with his life. Playfair says he 'wrote anxiously' to his friends in Edinburgh (none of these letters or their replies survive), but his unease may have been about much more than his career.

By the beginning of 1750 Hutton's son had been born. Where and when this happened or who the mother was, we do not know. There are no conclusive birth or baptism records and, although there is a James Hutton recorded in the tax records as living in London at the time, we cannot be sure it is the same man.[3] As for evidence of his son's early years, although we know that he was educated we don't know where or by whom. Few school rolls survive from that period and none lists a James Hutton as a pupil at the right time.[4] What we do know is that in his adult life the younger James Hutton lived at various addresses in the Wapping and Tower Hamlets districts in the city of London, close to the river Thames.[5] We know also that he worked as a clerk in the Post Office in Lombard Street.[6] In 1775, when he was in his mid-twenties, he married Alice Smeeton in London. But did he grow up there or move there as a young man?

Various theories have been suggested for Hutton being in London. That the mother of his child, born in Edinburgh or some-where else outside the capital, wanted to move to London to escape the scandal and that after his graduation Hutton visited her there to 'settle' her and make sure she and her son were properly provided for. But who was she and why did Hutton not marry her to make her respectable and legitimise his child? Was she already married to someone else? Was she a blood relative whom he could not marry? Did she die in childbirth – a not uncommon occurrence in the 18th century.

Or did he in fact marry her? In a later letter (see next chapter) Hutton writes to a friend 'but now [I] am e'en [even] wedded, and so must endeavour to restrain the wandering infidelities of the heart'.[7] It is not certain whether he means this literally – that he is already married – or metaphorically that he is 'wedded' to the long-term and all-consuming project he had by then embarked on.

If Hutton was in fact married, there are many candidates for his bride. An Elizabeth Galbraith married a James Hutton in Kippen, Stirlingshire, in 1747; Helen Chalmers married a James Hutton in Dunfermline in 1748. Further afield there were women married to men with the same name in Stirlingshire, Fife, Perthshire, Lanark and Norfolk.[8] In the same year a Helen Flint married a James Hutton in St Cuthbert's Kirk, Edinburgh, and later that year a Catherine Henderson married a James Hutton in the Canongate Kirk, Edinburgh. Intriguingly the fragments of the Canongate Kirk session minutes which survive for that period describe the 'interrogation' of a William Henderson and the description of a woman as 'with child'.[9] Disappointingly there is not enough of the badly damaged page left to be certain, but this could have been the merchant William Henderson questioned by the church minister and elders about his daughter Catherine (sometimes spelt Catharine). So the marriage soon afterwards could have been the result of the church and the bride's father forcing the groom to 'make an honest woman' of the expectant mother. If this was our James Hutton, was he recalled from Paris to be married?

This is circumstantial evidence, but is it convincing? We know there were two men named James Hutton living in Edinburgh at the time. The second was a merchant in Leith, supplying ropes and other goods to ships which used the port. But could there have been other men with the same name? Unfortunately, there could have been. Our James Hutton was born in 1726, but parish registers show one born in Edinburgh in 1722, another in 1727 and two in 1729.[10] All four might have been living in Edinburgh in 1748 and two of them might have been the husbands to Catherine Henderson and Helen Flint.

There are no records of similar weddings in London, either in the Church of England records, the minutes of the Scots Church which met in the Founders' Hall, Lothbury, in the city, or the surviving records of the non-conformist churches. But that does not mean no men named James Hutton were married in that period. Before Lord Hardwicke's Marriage Act of 1753 irregular or 'clandestine' weddings were common. The bride and groom needed only to give their consent to the union for it to be recognised. Clergy and witnesses were not necessary, though they were often present to provide proof that the marriage had taken place. The demand for clandestine marriages was met by institutions that considered themselves exempt from Church canon law and in some cases, like that of May Fair chapel, by a cleric who simply flouted the regulations. Prisons like the Fleet and the King's Bench Prison became popular destinations for couples interested in quick, no-questions-asked nuptials because of the number of clerics imprisoned for debt who had nothing to lose and welcomed the income. Many of them lived in the 'Rules' or 'Liberties', which were areas around the prison where prisoners could pay for the privilege of living outside the gates.[11]

In the summer of 1750, Playfair tells us, Hutton returned to Edinburgh, but for the next two years we know as little of what he was doing in the Scottish capital as we do about what he did in the English one. Did he leave London because the mother of his child rejected him? Or perhaps because he did not want to be encumbered with a dependent wife and son? A later episode in his life suggests that the opposite is the case, that he felt the absence of a family of his own and would have loved a son to guide and nurture (see Ch. 11).

Hutton was legally bound to Dr Young under the terms of his apprenticeship deed until April 1750, but he had been absent for at least half the five-year term in Paris, Leiden and London. Had he negotiated an early release? Or on his return to Edinburgh was he required to complete his indentured service by assisting Young, or working in his master's apothecary's shop? He was now a qualified doctor, although there is no evidence that he had ever treated a patient. In fact one of his friends later remarked: 'an attempt to consult him, or see him [on a medical matter] would have been met with a laugh, or some ludicrous fancy to turn off the subject.'[12] Working in the shop would have given him more opportunity to practise chemistry. Playfair says that one of the friends to whom Hutton had written was his university contemporary John Davie,[13] who shared his interest in chemistry. It is possible the two men conducted chemistry experiments together, but it seems unlikely that their plan to form a joint enterprise to manufacture industrial chemicals was formulated at this stage as Playfair claims – if it had been it would have made Hutton's next change of course even more surprising. He decided to become a farmer.

The Hutton family owned two farms in Berwickshire, about 35 miles from Edinburgh, less than a day's ride away. Nether Monynut, a hill farm of 590 acres (238 hectares) had been bought by his father William Hutton in 1710.[14] A few years later Slighhouses, an arable holding of 'four husband lands' – just over 100 acres or 42 hectares – was bought by his uncle John.[15] After the death of John it passed to William Hutton. James inherited both properties from his father. The two farms were about eight miles apart. Slighhouses also included 'houses, buildings, yeards, lofts, crofts annexis connexis, parts penticles and whole other pertinents', but there is no indication that any of the family had lived there, let alone worked the land. It is more likely that they were bought as financial investments, for the rent they would bring in. According to the land tax rolls for the period Monynut (or Monymill as it is called in the record) had an annual rental value of £141 13s 4d and Slighhouses £94 5s 5d). James Hutton may have visited the farms on his journeys to and from the south.

There was an increasing realisation in Scotland – still largely an agrarian economy – that the country's agriculture lagged behind other parts of the UK and the Continent in innovation and productivity. In 1723 the Honourable the Society of Improvers in the Knowledge of Agriculture in Scotland had been formed, and Allan Ramsay had written a poem in celebration in which he described Scotland as a 'nation long more keen of martial toil / than cultivating of a yielding soil'.[16] A number of progressive landowners – including Hutton's friends, the Hall family in Berwickshire and the Clerks in Midlothian – were actively improving their land and introducing modern farming techniques. In the 1740s William Cullen, who like Hutton had trained as a doctor and would later make significant contributions to medicine, included agriculture and agricultural chemistry in his lectures at the University of Glasgow.[17] Farming was considered with other sciences as part of 'natural philosophy' – a subject of intellectual as well as practical and financial interest. Certainly Hutton saw farming this way, writing many years later that he wanted to make 'philosophers of husbandmen and husbandmen of philosophers'.[18]

His interest had been sparked by reading Jethro Tull's book *Horse Hoeing Husbandry*, which had been published in 1733 and reprinted in several editions. It was not so much the techniques which Tull advocated – some of which were controversial and fiercely resisted by traditional farmers – but his willingness to challenge the old ways: 'I was charmed with Tull's system of pulverisation and cleaning the soil from weeds at the same time, being conversant with chemistry, I saw plainly his error with regard to the principles on which depended that fertility so important to the husbandman which Mr. Tull's system however beneficial in some respects is not properly calculated to produce.'[19]

During this time he was introduced to John Manning, four years his junior and just about to embark on a medical education, initially in Edinburgh and then in Leiden. The connection was probably made by Sir John Hall, Hutton's university friend, who besides having extensive farming and coal-mining interests in Berwickshire (including a farm close to Hutton's own) had business interests in Yarmouth[20] and Norfolk, where Manning's father Thomas was a farmer. It is likely that Hall asked Hutton to take the young Manning under his wing, help him find suitable lodgings and introduce him to people who might aid and advise him. (Twenty years later, Manning was to ask Hutton to provide the same services to the young Silas Neville.) Having told Manning of his interest in agriculture, the Norfolk man persuaded him that East Anglia was the place to see the most modern farming methods in action and recommended his father as the first point of contact.

In 1752 Hutton set off to learn about agriculture, mainly in Norfolk and Suffolk, but visiting other parts of the east of England and the agricultural areas of northern France and Flanders. He was to stay away two years. It is hard to resist the thought that he was going into voluntary exile – although from what we can only guess. The change of scene appears to have done him good because he enjoyed his study tour, although he was mostly observing and talking to farmers, rather than working the land himself. He spent his time 'in the most agreeable, if not the most useful manner'.[21]

After staying for a while with Thomas Manning, he spent a year with John Dibol[22] of Belton, near Yarmouth, probably as a paying guest. Hutton describes Dibol as 'a particular friend', and Playfair adds that he stayed in his house in comfort and benefited from the lessons he was given. 'He appears, indeed, to have enjoyed this situation very much: the simple and plain character of the society with which he mingled suited well with his own and the peasants of Norfolk would find nothing in the stranger to set them at a distance from him, or to make them treat him with reserve.'[23] He later moved to 'High Suffolk', the upland area where the heavier clay soils were more like those of Berwickshire. It was while in East Anglia that he furthered his interest in geology, making journeys to other parts of England, ostensibly to study farming. In a letter to Sir John Hall he described looking 'with anxious curiosity into every pit, or ditch, or bed of a river that fell in his way'.[24]

This sounds casual, but in a later letter to fellow rock and fossil collector Sir John Strange, Hutton showed an extensive knowledge of mineral deposits in England, from Northumberland to the south coast of England.[25] Describing the Isle of Wight Needles – chalk sea stacks – which were part of a 'ridge of indurated chalk', he added that he had followed the ridge from Corf Castle almost to Weymouth. In the same letter he mentions Portland, Bath, Oxford, Catton, Cambridgeshire, Lincolnshire, Derbyshire, Yorkshire and Northumberland and is knowledgeable about the geology of each place mentioned. We know that 20 years later he made journeys through the western counties of England and visited Wales, but his knowledge of the rest of England must have resulted from this time. Since he would have been travelling mostly on horseback, his research trips represented a considerable investment in time and effort.[26] He later boasted that 'I can undertake to tell from whence had come a specimen of gravel taken up anywhere, at least upon the east side of this island.'[27]

He finished his trip by sailing to Rotterdam (probably from Great Yarmouth) and travelling through the Low Countries of what are now the Netherlands, Belgium and northern France. Playfair says he was 'highly delighted' with the garden culture of farming he saw there, but as he was later to admit, not much of it was relevant to the much more challenging conditions he was to find on his own farms. He returned to Scotland via London in the summer of 1754, but even then, brimming with theoretical agricultural knowledge, he hesitated before committing himself to what was to prove to be a lonely and arduous life as a farmer.

Chapter 6

I shall die like a cock . . .

EXCEPT FOR HIS two years in East Anglia and the Low Countries, Hutton had lived all his life in cities. He was gregarious and enjoyed the company of friends. His move to Berwickshire took him into a very different world. His farm at Slighhouses, where he now made his home, was in the parish of Buncle and Preston – ten square miles with a population of less than 700. In contrast Edinburgh, one of the most crowded cities in Europe, had 30,000 people living in a tenth of the space.[1] His nearest neighbours were no longer next door, but he wasn't entirely without friends, even though they were now more distant. Sir John Hall's estate was at Dunglass, to the north-east, but that was three to four hours on horseback. The family of James Pringle, another university friend, lived at Stichill to the west, but that was even further, and the Clerk family were a day's travel away.

His studies in East Anglia and the Continent had given him ideas about modern farming methods and technology but had not prepared him for the primitive conditions he would find on his own land. He was fit and still under 30, but he had not been used to manual work. He had labourers working for him, but still had to spend long days directing and working alongside them and over-coming their mistrust of the new practices he wanted them to adopt. The land was still divided into 'rigs', domed strips 10–20 feet (3–6 metres) wide and perhaps hundreds of yards (metres) long, with each rig separated by a ditch. Heavy rain washed the topsoil from the crown into the ditches. Hutton saw that the fertility of the land was literally draining away and he wanted to enclose it into flat fields separated by stone walls.

Ditches around the perimeter would enable him to drain the land without losing the soil and to contain sheep or cattle to manure the ground between crops. The task was laborious and the equipment he had was heavy and basic. For moving stones to build walls 'I'm obliged to use a great clumsy wooden slipe* and drags the whin stones thro' the rough land (sore against their will in as heathenish a manner as if I had never seen the Gospel).'[2] To flatten the ridges of the rigs took 20 ploughings,[3] with a heavy wooden plough pulled by a team of six or eight oxen, which needed a ploughman to try to keep the blade on a straight path and a driver to lead the beasts. Sometimes there had to be a third man to remove large stones so that they did not damage the plough.[4] It was hard, slow work and even then some of the ridges needed to have earth shovelled from the crown to the hollows.

Fatigue and loneliness quickly wore him down and he began to have doubts about his decision to go to Slighhouses. He made desperate appeals to his friends begging them to visit him: 'Dear Sir, will you come and see, but do not flatter my expectations with a vain hope,' and 'when will Tam Sharp or you come out a jaunt? 'tis an easy day's ride.'

A few months after he arrived he wrote to George Clerk, brother of John, his fellow anatomy student. The extraordinary letter, in which Hutton bares his soul, shows how close he had become to the Clerk family, treating George, who was 11 years older, like an elder brother and promising a separate letter to Dorothea Clerk Maxwell, George's wife. In a tone which lurches from whimsy to despair, he admits to missing the easier life of Norfolk – when he was an interested observer rather than a practical farmer who had to make his ideas work – and how if he could 'disentangle his foot' he would go back there. But he had made his bed and accepted he had to lie in it. 'Lord what do it signify vexing plaguing and perplexing oneself thus in a cursed country where one has to shape everything out of a block and to block everything out of a rock.'[5] In Edinburgh, he adds, he could commission a builder or order wood and other supplies to be delivered – 'and then I goes me out about my business in peace and lays me down to quiet rest, but here, I may sleep on and dream to all eternity without advancing one step. My houses

---

* A wooden platform or sledge without wheels, dragged by men or oxen, used for moving heavy or cumbersome loads.

are not a foot more forward than the day I came to this country.'
Isolation brought him close to despondency:

> I shall soon forget to both read and write. You may laugh at
> Ovid and his fables as you please, but I find myself already
> more than half transformed into a brute beast. There is noth-
> ing of the Christian left about me except some practice of piety
> and prayer. O, as for the last, I shan't need to trust in it, so long
> as I have so much subject matter; I ain't like in haste to wax too
> fat, nor fart, nor flung neither – but faith, faith of all things, is
> what I want most; I haven't a single grain of it to do me any
> good.[6]

Hutton looked back with nostalgia to rural life in Norfolk and,
according to Playfair, he often described 'with singular vivacity, the
rural sports and little adventures, which, in the intervals of labour,
formed the amusement of their society'.[7] His neighbours in
Berwickshire did their best to invite him to their social gatherings,
but his newly adopted cynicism did not allow him to enjoy them:
'They had me at a feast of Baal [at Eyemouth] where was a sow, an
honest sow, roasted i' the guts, so we had a dish of surprised pig and
I did eat thereof. They also led me up in the dance, but I will enter
no more into their high places.'

   The magnitude of what he had done in cutting himself off from
his friends had dawned on him: 'O it would grieve your heart . . . to
hear what a lone [life I lead].' There was a spark of pleasure in the
news of the birth of a calf to one of his cows: 'Give me joy there is
born into my family a male child – and the mother is a fair way of
recovery' – but then despondency again and a broad hint that his
troubled emotional state was because of the failure of his own relationship and that he had been rejected, rather than abandoning the
mother of his son: 'O if the ladies were but capable of loving us men
with half the affection that I have towards the cows and calfies that
happen to be under my nurture and admonition, what a happy
world we should have! But their minds are wholly set on vanity –
adieu.' And then, although some words are missing because of
damage to the paper, a painful cry from the heart: 'Ay lost for
_____ to _____ friends and what is worse, they are irrecoverably

lost I doubt to me – there is no truth in proverbs, for here's a great loss without the least shadow of gain that I can see.'

His heart was broken. He had been able to forget his estrangement from his baby son and the boy's mother while he travelled, but now he had little to occupy his mind when the day's hard labour was over. And there is a suggestion he had suffered a new rebuff, which prompted a jaundiced view of the female sex. In his letter to George Clerk he hints that he has been slighted by a woman for a second time:

I won't let any of the fair kind of creatures know of my distresses, it would but kittle the malicious corner of their heart to hear the afflictions of a hardened wretch whom they could never make to groan – fine talking of the Enemy says you, when at a distance, ay, so it is. Here to be sure I have it all my own way, but at any time, you know, one had as good give the slight as take, so I served Miss Pumphry you may remember, hang 'er for a brimestone [sic], I believe I'm as well quit yet I should have thought my line to have fallen in pleasant places with her, for all that.

We can only guess at the identity of 'Miss Pumphry'. As so often in Hutton's life, conclusive evidence is absent, but a series of letters between Sir John Hall and his relative Sir John Pringle could refer to Hutton and a possible reason for a failed relationship. Hall asked for help for 'a friend' who was having an emotional problem with a woman (who is never named) and Pringle agreed to meet her. In the final letter in the exchange, Pringle, writing from London on 23 May 1758, says:

I believe the affair is off, without any quarrel between the parties. The lady behaved most handsomely, she came because she promised to come & because many things could be spoken that could not be written, that her headaches for some weeks past had returned with such violence and frequency that she could not [enter] in honour into that state which must make her a burden to the person she wished the best to. That for some time before she had flattered herself that her constitution had undergone a favourable change but that she found she was mistaken & was even the worst for the journey which otherwise she hoped might have been for her advantage.

Let me add that the lady never gave any absolute promises, but always reserved a liberty to both to draw back if either found it expedient to do so upon meeting. To all this declaration nothing could be objected, but on the contrary, the candour and generosity of the behaviour was to be highly praised. I have a notion that she & her companion will set out next Friday for the north and that they will take it in their way. This, I say, is the scheme at present, but I cannot say what changes there may be made by the chapter of accidents. You see here much I have to praise in your reticence in this affair. Whatever be the event I beg you would continue to show your regard for her as she truly deserves it.[8]

Hutton may also have had hopes of a relationship with Magdalene Pringle, sister of his university friend James. Known as Maudie, she was 26 in 1755 and a noted beauty. Hutton could have met her at her family home at Stichill or at the estate at Dunglass of her cousin Sir John Hall. Two letters from her to Sir John refer to 'a present from the doctor', and 'a letter from the doctor',[9] which may be references to Hutton.[10] But unknown to him her affections lay elsewhere – she was in love with Sir John. They kept their affair secret out of fear that their families would disapprove of a marriage between cousins, but eventually overcame any objections and married in 1759. In 1761 Maudie gave birth to a son, James (named for his grandfather, rather than Hutton), but she died three years later.*

In another letter from Hutton to John Bell, another of his university friends and now his solicitor, he hints again of his low emotional state, partly brought on by exhaustion after throwing himself into work: 'it is not in the nature of things to work all day and write all night otherwise you might have heard from me ere now ... Everything conspires to break my heart, but I shall die hard John, I

---

* In adult life the boy, later Sir James Hall, became a close friend and scientific collaborator of Hutton. See later chapters.

shall die like a cock, tho' at present I assure ye I live like a Capon'[11] – that is like a neutered male.

To add to his troubles, Bell told him that one of his tenants in an apartment left to him by his father was not paying rent. Hutton was reluctant to go to court, which meant delay, uncertainty and cost, but he relied for his income on rents and interest payments from loans he had made. The premises were at Gosford Close, which had been left to the family by his father. Records for 1752 show that Mrs Hutton (presumably James' mother) was living at Fishmarket Close. Three further properties are shown belonging to William Hutton, merchant. These could also be the family's assets, still in the name of James' father.[12] Slighhouses was costing James Hutton money, rather than providing a living, and the modest income from lettings had to support his mother and his unmarried sisters as well as himself. The sums involved were small, but there was no quick resolution – the case dragged on for more than ten years. Bell obtained a Sheriff Court judgement against the tenant, goldsmith William Livingstone, but he died in 1756 with arrears outstanding and his widow was no better at paying. Hutton obtained a warrant of sequestration from the sheriff over Mrs Livingstone's effects, but she appealed to the Court of Session. Bell told the court she had been 'clandestinely' removing her things from the house and 'disposing of them at pleasure'.[13]

In 1756, two years after he moved to Slighhouses, Hutton suffered a further blow. His mother Sarah died. After his father's death when James was three years old she had been his guide and protector. She had stood by him when he abandoned the legal career she wanted for him, and she provided support during the social and emotional crisis over the birth of his son. We do not know if he went back to Edinburgh to join his sisters for her funeral. She was buried in the Balfour family tomb in Greyfriars churchyard, alongside her infant sons, William, who had died at the age of two, and John, who died aged three.

Despite his problems, little by little Hutton brought the farm under control and was able to put into practice some of the techniques he had seen on his study trip. On his return from England he had brought with him a Norfolk plough which, being made of steel rather than wood, was lighter and more easily manoeuvred. It

was more efficient than the traditional Scottish plough and could
be pulled by a pair of horses, controlled by the ploughman – no
need for the brute power of the oxen, or a second man to lead them.
It could plough a larger area in a shorter time – an acre in a day –
but it was viewed with scepticism and amusement by his farm
workers and neighbours, who believed that such a toy could never
work in the heavier soils of the Scottish Borders. Hutton wrote to
his friends in Norfolk asking them to hire him a ploughman, but no
one was willing to face 'banishment' to Scotland without meeting
the man who was to employ them. Hutton had to go back to East
Anglia himself, where he found an unemployed ploughman who
agreed to come north.

He proved his worth and the detractors wrong: 'I had occasion to
put this instrument to the most severe trial that perhaps can be
made. This plough with a pointed share broke up land possessing
in a high degree almost every possible difficulty. It was a field which
at that time did not deserve to be called arable land; and afforded a
remarkable specimen of ploughmanship as well as trial of the
plough. But the ploughman was a Norfolk man and master of his
business.'[14] The man could work the plough, but Hutton's hope that
he might in time become a farm manager or at least be able to train
others proved unfounded; he wasn't up to it.[15] The ploughman
stayed at Slighhouses until 1763 and Hutton was sufficiently
impressed with his work to invite his neighbouring farmers to
ploughing demonstrations.[16]

With the fields enclosed and ploughed, Hutton set about improving the land by dressing the soil with fertiliser. He experimented
with salt, seaweed (raw and burnt), muck (animal manure from his
byres), marl and coal ash.[17] He monitored the results carefully by
measuring the crop against what he might have expected from
unfertilised fields. All but muck and marl he abandoned as not
worth the effort. The marl – soft rock and soil rich in lime (calcium
carbonate) – he dug out from his own land and made such use of it
that by the time he finished farming he had formed a quarry 100
yards long, forty wide and five deep.[18] He first used it indiscrim-
inately – 400 two-horse loads per acre – but quickly discovered it
was not all of the same quality:

The marl was carted and dug without respect for what was good or not ... When the corn came up, I did not perceive the effect of this when in the field, but happening to look down from the hill ... I saw the field in some places chequered with the deepest and the palest green. I immediately went into the field and examined the marl in those spots and then examined the strata in the marl pit; this put the matter out of doubt.[19]

Examining the strata was an early example of Hutton's growing interest in geology, although he did not call it that and saw no distinction with his interest in fertility and soils. Here in front of him was a graphic illustration of his farming problem: the rock had been laid down in layers, some richer in lime than others. But differentiating between the two while the marl was being dug and carted was more difficult. To overcome the problem Hutton later devised a practical test to gauge the quality of the marl.

As a scientist, he was not content merely to observe the benefits of feeding the crop, he also wanted to understand the processes. How was it, for example, that two substances as different as muck and marl could produce the same effect?[20] He was also interested in how exhausted or semi-exhausted land could be brought back into fertile production. He advocated planting buckwheat as a green crop which could be ploughed back into the soil (with a modified plough) to provide organic enrichment, or under-sowing barley with grasses and clover so that after the corn harvest they would provide pasture for cattle or sheep, which would manure the land as they ate.[21]

His curiosity did not stop with soil fertility. He was interested in the influence of light and tried growing carrots and radishes in the dark. They grew to maturity, but had no colour, taste or smell. He wanted to know how the weather affected the crop and tabulated the effect of climate on plant growth.[22] For his experiments with the effect of warmth on crops he invented a thermometer contained in an insulated wooden box, as a simple way of measuring average temperatures. He tried out different crop rotations. On the suggestion of his Norfolk ploughman he adopted mass planting of turnips to counter the effect of parasitic fly damage and he tested various methods of treating wheat to reduce the damage from disease.

Hutton practised mixed farming at Slighhouses – he grew cash crops such as barley, planted pasture for sheep and cattle and grew root crops like turnip, which could be eaten or used as animal feed. It is also clear from his later writings that the farm produced milk, butter and cream. Hutton had observed dairying in Norfolk, but on his own farm he adopted Devonshire ways of eliminating the unpleasant taste of milk from cows fed on turnips. This involved 'scalding' the milk before making it into cream or butter. He admitted it too gave a distinctive taste, but 'this peculiar taste is not, like that from the turnip, disagreeable or generally obnoxious to the palate; it is rather an agreeable taste, and to those who are not prejudiced against it, preferred to that which is made from raw cream.'[23]

Hutton believed sheep were better adapted to his land than cattle. Not only did they give a quick profit – in the form of a fleece and a lamb every year – but a longer-term benefit in manuring the land with less effort than with cattle. He believed that ruminant animals not only needed to be well fed but should also be given the time to properly chew the cud, 'during which operation they should not be disturbed'.[24] Nowhere in Hutton's letters or later writing does he mention his second farm at Nether Monynut. It is possible it remained let to someone else to provide rental income.

He was by no means the only improving landowner. Although he pioneered some innovations such as the steel plough, which was widely adopted in Scotland, many of his near neighbours were also experimenting with new crops and methods of increasing productivity. These included his close friends Sir John Hall and George Clerk on their own estates. Further afield the judges Lord Kames, Lord Braxfield and Lord Monboddo were all active improving farmers – Kames later publishing his own book: *The Gentleman farmer, being an attempt to improve agriculture by subjecting it to the test of rational principles (1776)*.

Time healed Hutton's emotional wounds and the progress he made with the farm enabled him to recover his strength. A second light-hearted letter to George Clerk, written in May 1757, two years after he arrived at Slighhouses, makes no mention of women or loneliness and shows that he had been taking time out from the farm to visit friends and had kept up his interest in geology. 'You'll have no doubt given me up long ago for lost, indeed I have been so and found again several times since I had the pleasure of chaping* the stones on the head with you.'[25] He adds that a list of his wanderings would resemble 'a maze more formidable than the labyrinth of Crete'.

After his marriage to his cousin Dorothea,[26] George moved to one of the Clerk family estates at Dumcrieff, near Moffat. Like Hutton, he had studied at Edinburgh and Leiden and shared with him an interest in science. He wanted to develop the potential of his holdings. It appears from the letter that he asked Hutton to inspect land at Leadhills, South Lanarkshire, possibly with a view to opening a lime quarry. Hutton had to disappoint him – he found no limestone, but 'a clew of spar' led him to the 'metals by the nose'. The spar was probably crystalline minerals, quartz or feldspar. The area had been mining lead and zinc for hundreds of years. The rest of the letter is a tale of misfortune told with good humour – of nearly drowning himself and his horse, Molly Meg, by taking a shortcut through a ford, of getting lost in the dark ('cold and wet I was and thought to have died the death that night') and finally accompanying two friends from Norfolk on the first leg of their journey home from Edinburgh, 'to the great scarification of my flesh apon an Edinburgh hack†.' He treated his saddle sores with a 'Ladies court plaister‡, which I was obliged to use a little rudely'.

The trip was the precursor of a tour of central and north-east Scotland Hutton made with George Clerk in 1764. Clerk had been appointed as one of the commissioners of the Forfeited and Annexed Estates – land seized from lairds who had supported the Jacobite rising in 1745–6. Commissioners were unpaid and were expected to visit and assess the value of the land, rents and other income and make recommendations about how the proceeds should be spent – on housing, education, manufacturing, improving agriculture and fisheries and on roads and canals. Their journey, which took in Crieff, Dalwhinnie, Fort Augustus, Inverness, Easter Ross and Caithness, returning to Edinburgh along the coast via Aberdeen,

---

* A Scots word for hitting or striking
† A rented horse
‡ Sticking-plaster made of silk (black, flesh-coloured, or white) coated with isinglass, used for covering superficial cuts and wounds

must have taken them weeks, if not months. Clerk would have valued Hutton's opinion on the quality of each of the farms they visited and assessed, and perhaps in giving a second opinion on surveyors' reports he commissioned. Hutton also analysed a sample of marl from an estate in Perthshire and advised Captain Lockhart, whose estate was at Balnagowan, on the Cromarty Firth, to use shells as fertiliser.[27] But from later letters Hutton wrote to a fossil collector it is clear that his personal interest was turning more and more to geology, and he used the opportunity to increase the extent of his mental geological map of Britain.[28]

Hutton's experience as a farmer and his observations of rock formations and erosion stimulated his thinking about the processes by which the Earth had been formed. It would be more than two decades before he was ready to publish his paper on the age of the Earth, but from papers found after his death it was clear he had begun to work out his theories while he was still at Slighhouses.

Shortly after returning from his journey with George Clerk, Hutton gave up farming. He was disappointed in the ploughman he brought from Norfolk: 'He remained with me several years; but as he neither could manage the business of my farm, nor could act properly under the management of another, I gave up farming, just at the period when the land was brought to such a state as it might have been cultivated with pleasure and to good purpose.'[29] But he maintained his interest in agriculture even after his return to Edinburgh: 'In conversing with a very intelligent brewer in this City, I found he had a kind of barley from Norfolk which he commended much, altho' he had barley of a larger size from East Lothian; but this Norfolk barley was a beautiful grain and malted very kindly. I begged to have a boll of that barley, which I then sent out to my farm in Berwickshire. I there multiplied it and it prospered very well.' The final outcome, however, was not good: 'that which is a virtue for the sheltered lands of England, may be a serious vice in the corn for the open fields of Scotland situated in such a windy region'.[30] Late in his life he also started to work on a treatise on agriculture, but it remained unpublished at his death.

Having made the farm more productive Hutton had no problem in finding a tenant – and he had other adventures awaiting him in Edinburgh.

Chapter 7

Turning soot into gold

WHILE HUTTON WAS farming at a loss, his university friend John Davie was making money. When Hutton had started his short-lived legal apprenticeship in 1743, Davie had become apprenticed for three years to John Walliburton,[1] described as a merchant. We don't know exactly what business he was in, but during the next twenty years Davie had become prosperous enough to start accumulating land. In 1754 he had taken a 'tack' – a ground lease – on a piece of land just south of the city walls of Edinburgh. His landlord was Dame Elizabeth Carnegie, widow of Sir James Nicolson. Davie's holding was separated from the 'park' of her mansion, Nicolson House, by a wooden fence, which Davie was required to maintain for 19 years. He paid three guineas (£3.15) a year, which gave him the 'liberty to quarry for stones and build houses thereon',[2] although it is questionable that he ever did.

Over the next decade Edinburgh started to expand to the south and developers such as James Brown, who built Brown's Square and then the larger George Square, and the architect Robert Adam, who built Adam Square*, began to buy up land for superior houses. The area had a dual appeal – in contrast to the old medieval city, there was space to build larger homes intended for single families, either with private gardens or facing landscaped squares. Secondly, it was outside the city's 'royalty', and not subject to the council's taxes. In 1762 Lady Nicolson – 'observing the amazing spirit of building which has gone forth in the metropolis' – took advantage of rising prices to 'feu' (grant leases) on ten acres of her grounds.

* Brown's Square and Adam Square were on the site of the present Chambers Street, Old College and the National Museum of Scotland.

She advertised in newspapers and had a plan engraved and printed, which was pinned up on the walls of coffee houses. The land was divided into individual plots large enough for one spacious house and there was to be a wide new road running north-south (to be called Nicolson Street) and a crossroad running east-west. There appears to have been high demand and she quickly found takers for all the plots and later 'feued' the whole park in 19 lots. Davie bought three, giving him a substantial area of land, described as '5 roods, 17 perches and nine yards in English measure'.[3] In modern terms that was one and a half acres (0.6 hectare).

However, a dispute developed between Lady Nicolson and her new leaseholders over her refusal to build the east-west road, which they considered necessary to access their plots. Failing to reach agreement, the 'feuers', led by John Davie, took the case to court and in a submission admitted that although they had paid highly for their leases, 'it will not be denied that they could have sold them for double the price'. Failure to build the access road would severely reduce the value of the holdings. To try to facilitate a settlement, Davie offered to surrender his 'tack' over the original piece of land he had leased since 1754. As a gesture of goodwill, he offered this at no cost to Lady Nicolson, suggesting that he cannot have built either houses or a factory on the land. Certainly, William Edgar's map of 1765 shows nothing that looks like a factory on the ground.[4] The court papers do not record whether a settlement was reached, but Lady Nicolson died the following year. The road was built, so either she must have agreed or it was constructed by her heirs.[*]

However, that was not the end of Davie's property ambitions. In 1766 he acquired an estate at West Calder, 20 miles west of Edinburgh. John Bell, who was Davie's lawyer as well as acting for Hutton, arranged the purchase from Davie's father-in-law, James Flint of Gavieside, who had also been an Edinburgh merchant and member of the council. Davie was married to James' daughter Mary and the couple went on to have 12 children. The sale included 'lands and barony of Marjoribanks, viz and all and whole of the lands of Easter Blackmyre, Killingdean and Brothertown' and included the 'coals, coalheughs, milns [mills], woods, fishings,

---

[*] It is now West Richmond Street.

mosses, muirs, meadow ponds, pendicles and pertinents'. The king was nominally superior of the estate and the deeds required Davie to pay 'blench duty' of 'one pair of gilt spurs or two shillings silver' every Whitsunday. Blench duty was an archaic theoretical obligation and Davie had to stump up 'only if asked'.[5] Henceforth he could style himself 'John Davie of Gavieside', adding a veneer of landowning respectability to his wealth earned in trade.

We don't know precisely when James Hutton returned to Edinburgh. Playfair says it was 1768,[6] but it was almost certainly earlier, perhaps late in 1764 or 1765. He had been interested in the applications of chemistry in agriculture and manufacturing, particularly the textile industry, which was expanding rapidly. He was possibly encouraged by George Clerk, who had built a linen factory on his estate and written papers on the treatment of woollens for the board for the improvement of manufacturing and fisheries. As prosperity increased there was a demand for coloured cloths, but most dyes had to be imported and were expensive. Entrepreneurs looked for ways of using local materials and reducing the cost. In 1764 a factory had been established in Leith for producing red and purple dyes from lichens and several dyeworks had been established in Glasgow.[7] Hutton advocated planting woad (*Isatis tinctoria*) as a substitute for indigo, a blue dye which could only be produced in the tropics, although he admitted he had no personal experience of growing it. But he had tried growing madder (*Rubia tinctorum*) and had experimented with boiling samples of cloth in different intensities of the solution – giving a range of brown and red colours.[8]

Whether Hutton and Davie considered setting up a dyeworks we do not know, but they were interested in industrial chemicals, particularly in sal ammoniac (ammonium chloride, NH_4Cl), which had wide applications in the metal industries, in dyeing and textiles, in baking and as a food flavouring. The white crystals occur naturally but are rare, so most were produced artificially and had to be imported. Who first had the idea to try to make it in Edinburgh is not certain, but given Hutton's interest in chemistry and his education, it is probable that he provided the scientific knowledge and technical innovation and that Davie provided the premises, the business acumen and perhaps the working capital. It is unlikely that

Hutton, still living on his inheritance from his father and with three spinster sisters to support, had much capital to contribute. In 1765 they went into partnership – Davie and Hutton – to manufacture the substance in a factory on Davie's land at Newington.[9]

Sal ammoniac had a long and exotic history. It was exported from the wastelands of Central Asia to China and the Arabic world and was used in the preparation of drugs and medical elixirs. It was traded (and perhaps manufactured) by the Venetians and imported into Europe from Egypt, where it was said to be made from the 'sublimation' – the heating of a solid to turn it into a gas – of soot from furnaces fired with dried camel dung. If Hutton had attended lectures at Le Jardin du Roi he had possibly heard Rouelle describe sal ammoniac being found in volcanic regions and claim that he had been able to make it using 'vitriolic acid' (sulphuric acid) and peat.[10] In 1720 other French chemists had published an account of the production of sal ammoniac, but there was controversy over whether it was possible to produce it from soot alone, or whether other substances needed to be added, such as urine or salt. In 1735 Henri-Louis Duhamel du Monceau had published a paper examining the various competing processes,[11] which were also described in a chemistry textbook revised by Rouelle.[12]

Boerhaave's textbook on chemistry, published in Latin in 1732, also described several methods of making sal ammoniac, including from soot. It is possible Hutton read the book in the library of the University of Leiden during his month there in 1749.

Most of these processes used soot from furnaces burning animal dung or bedding straw which had been soaked in urine, which would have provided the ammonia needed in the process. Hutton and Davie proposed to manufacture it from coal soot, of which Edinburgh – known as 'Auld Reekie' because of its thousands of smoking chimneys – had an inexhaustible supply. The city was surrounded by coalfields in the Lothians and across the River Forth in Fife, and cart-loads of coal were brought into the city daily and unloaded from ships at the port of Leith. Coal was essential for heating and cooking.

Hutton, like Newton, for all his rationality and belief in science, still had a lingering interest in alchemy and the belief that the Philosopher's Stone – a mythical substance said to be able to turn

base metal into gold – might still be found.[13] Rouelle also covered
alchemy in his course: he did not doubt the word of the great men
who claimed to have been able to produce gold from lesser metals,
but he wanted to see it done for himself before he would be
convinced.[14] What Hutton was proposing was the 18th-century
industrial equivalent – transforming worthless black powder into
white crystals which could be sold at a profit. But could sal ammoniac be made only from coal soot, which is mostly amorphous
carbon?

Boerhaave had mentioned soot from 'a vegetable origin' and its
difference from coal soot in his book. If Hutton had found a way of
manufacturing sal ammoniac from coal soot, it was a scientific
breakthrough which had never before been achieved.

Some of the most eminent chemists of the day were sceptical,
including Joseph Black, still at this time a professor at the University
of Glasgow, but soon to move to Edinburgh as professor of medicine and chemistry. Black and the chemist and author James Tytler
thought that sulphuric acid and common salt would need to be
added to the soot,[15] but neither man was able to substantiate his
claim because Hutton and Davie kept the process secret and were
very reluctant to allow visitors into the factory. Playfair believed
that the process only required soot; modern chemistry accepts this
is possible, although vast quantities of soot are needed to produce
the white crystals in commercial quantities.

To secure a supply of the raw material Hutton and Davie made
an agreement with the Society of Edinburgh Chimney Sweeps,
known as the 'Tron men' because their office was in the guard
house near the Tron Kirk in the High Street. Despite the secrecy,
an eye-witness account of the process inside the Davie and Hutton
factory was published: 'Globular glass vessels, about a foot in
diameter, are filled to within a few inches of their mouth with
[soot], and are then arranged in an oblong furnace, where they are
exposed to a heat gradually increased. The upper part of the glass
balloon stands out of the furnace and is kept relatively cool by the
air. On cooling, the upper parts of the globes are found to be lined
with sal ammoniac in hemispherical lumps, about 2½ inches thick.
26lbs. of soot yield 6 of sal ammoniac.'[16] (In metric units, 11.8kg
yielded 2.7kg.)

To get the crystals out of the glass vessels probably meant breaking them, which would have added to the expense. To reduce the cost it is likely that the containers used by Hutton and Davie were old sulphuric acid bottles from the Prestonpans Vitriol Works, founded by Dr John Roebuck and Samuel Garbett in 1749 and only a few miles from Edinburgh.[17] Roebuck, like Hutton, had studied medicine at Leiden and Edinburgh, where he had taken the chemistry course of Professor Pummer.[18] Vitriol (sulphuric acid) was used in bleaching linen cloth.

The Hutton-Davie venture appears to have been a success from the start and grew to occupy a site of half an acre between Nicolson Street and the present Davie Street.[19] With his new country estate, Hutton's partner may have called himself 'Davie of Gavieside', but to the citizens of Edinburgh he would be known henceforward as 'Sootie Davie'. The business continued for nearly 40 years, but as soon as it was in production and the scientific and practical challenges had been overcome, Hutton lost interest and was enticed into another, even bigger challenge. With Davie now running the factory, the partnership continued to provide an income for both men. It did not make Hutton wealthy, but it allowed him to indulge his passions for science and engineering. He was never idle, but he did not have to work to earn a living.

Production of sal ammoniac from coal carried on at least until 1783 when a new, cheaper source may have become available. Joseph Black, who by this time was well established as a professor in Edinburgh and had become a close friend of Hutton, was asked to provide an assessment of the tar works established by Lord Dundonald at Culross, on the Fife coast of the Firth of Forth. Dundonald had started his career, as Archibald Cochrane, in the army and the navy and inherited the title from his father in 1778. With it came estates in Fife and Midlothian, a house in London and substantial debts. An inveterate inventor, he determined to use his talent for innovation and his naval experience to restore the family fortunes. While in the navy he had seen the ravages caused to wooden ships by sea worm and the high price that ship owners had to pay to protect the hulls with tar and pitch imported from Scandinavia – where they were produced from wood.

Dundonald believed he could make tar much more cheaply from coal, which he had available from his estate at Culross. This was not a new idea, but in previous attempts coal in closed containers had been heated by external coal fires – thus using twice as much coal and rendering the process uneconomic. Dundonald's insight was to see that he could burn the coal directly. In 1781 he built 20 brick stoves in which the coal was burnt with as little air as possible and the fumes piped into a brick tunnel 100 yards long and supported on brick arches. At the end the smoke was disgorged into a shallow pond. Bitumen was condensed to form tar, which could then be refined into various products for a wide variety of uses from protecting ship's hulls to making Japan black lacquer for fine furniture. From 120 tons of coal, Dundonald produced 3.5 tons of tar (122 tonnes yields 3.55 tonnes) – with the residue 'cinders' sold as coking fuel for metal refining.[20]

But enterprising though he was, Dundonald was a poor businessman. He had borrowed to establish his business, which produced profits more slowly than he had expected and one of his creditors was threatening to force him into bankruptcy. To save the firm he formed the British Tar Company and brought in external shareholders. Before they invested, they asked Joseph Black to visit Culross and assess the process and had Adam Smith look over the accounts and the financial projections. Black's report was favourable. He calculated that the business could be solidly profitable in peace time, but in wartime with increased demand for ships, it would make twice as much money. He also saw an opportunity that Dundonald had missed. One of the by-products of the process was 'volatile alkali' (ammonia, NH_3), which was condensed in the water of the final pond and could be used in the production of sal ammoniac. Black sent a sample to Davie and Hutton, who made Dundonald an offer.[21]

This would have increased the revenue of the Culross factory, but Dundonald did not immediately accept it and investigated his own methods of manufacturing sal ammoniac. We don't know whether a deal was eventually made by which the Davie and Hutton partnership moved from being a manufacturer of sal ammoniac to buying it wholesale and selling it on to their customers. It is possible that Dundonald turned down the offer and that Davie and Hutton continued to manufacture the chemical from soot. But despite the success of Dundonald's process he did not prosper. He fell out with some of his shareholders and failed to convince the Admiralty to use tar on British naval ships rather than sheathing their hulls in copper, which was their more expensive preferred solution. He went on to develop other chemical processes and published scientific papers, but died in Paris in 1831 in poverty. His oldest son, Thomas, who became one of Britain's most successful admirals during the Napoleonic Wars, lamented: 'His discoveries, now of national utility, ruined him and deprived his posterity of their remaining paternal inheritance.'[22]

The same fate did not befall Hutton and Davie. Despite some competition from Glasgow, the Newington works appears to have continued in business. After Hutton's death it was carried on in Davie's name alone,[23] but it appears to have closed after Davie died in 1803.

Chapter 8

Hutton dreaded nothing
so much as ignorance

THE EDINBURGH TO which Hutton returned after his self-imposed rural exile was a fast-changing city. The squalor of the medieval Old Town was giving way to new open squares and larger houses. By the 1760s there were new developments to the south of the city walls – Argyle Square, Adam Square, Brown's Square and, the largest of them all (and the only one which survives today), George Square. In the centre of the city, the visionary Lord Provost George Drummond, who had secured the first purpose-built infirmary in 1741, had demolished a row of unsafe tenements opposite St Giles' Cathedral and was building the Exchange, an elegant new commercial heart designed by Robert Adam. Drummond had also been the political force behind the publication of proposals for extending Edinburgh to the north, by draining the Nor' Loch and building a new town on the land beyond. As a first step he planned the construction of a huge new bridge (North Bridge) to join the old to the new, but died before it was completed.

The Scottish economy was recovering from decades of stagnation or slow growth. Agricultural prices were increasing, encouraging landowners to improve their estates and increase their rents. The linen industry had been given a boost by the launch of the British Linen Company in 1746, based in Edinburgh. The manufacturing industries – textiles, brewing, tanning, metal-working, printing – were growing, but it was increasingly a city of service industries and the knowledge economy – law and public administration, banking and religious and intellectual thought, literature and learning.

The philosopher David Hume, famous (or in some circles infamous) following the publication of his books and essays on human understanding, behaviour, morals and religion, had returned to Edinburgh after years in France and England. He had been rejected for the chair of moral philosophy at Edinburgh University because of his atheism (and was to be rejected for the chair of logic at the University of Glasgow for the same reason), but was made keeper of the Advocates' Library in 1752 and began a productive and successful period of writing and publication. Adam Smith, although yet to publish his major books on moral philosophy and economics, gave a series of public lectures at Edinburgh University between 1748 and 1751 on rhetoric, the history of philosophy and jurisprudence.

The foundation of the Select Society in 1754 brought together the intellectuals and the newly wealthy gentry and nobility, 'who wished to acquire polish from consorting with the literati as much as the literati desired status from the patronage of the social elite', says the historian T.C. Smout.[1] The Philosophical Society, which had gone into decline following the death of the mathematician Colin Maclaurin in 1746, had been revived by a group which included Sir John Clerk, father of Hutton's friends John and George Clerk.[2] In these societies landowners, mine owners, manufacturers and merchants discussed the issues of the day with scientists, mathematicians, lawyers and men of literature and religion. At the centre of this intellectual debate, the university was attracting new talent. When Andrew Plummer died in 1756, William Cullen, doctor and chemist, had been lured from the University of Glasgow to take his place as professor of chemistry, and ten years later Cullen's protégé Joseph Black followed the same path east to take the chair of medicine and chemistry.

Black had been born in 1728 in Bordeaux, where his father worked in the wine trade. After school in Belfast (his father's home city) he attended Glasgow University to study medicine, but took his degree in Edinburgh in 1752. His dissertation was on his discovery of 'fixed air', which we now know as carbon dioxide. Later he proposed the concepts of latent heat and specific heat to explain certain physical phenomena. Both terms are now key to our understanding of the physics of heat. It is possible that Hutton met Black at this time, but it would have been a brief acquaintance: Hutton left soon afterwards on his agricultural study tour and Black returned to Glasgow to work in the university. By the time he returned to the capital, Hutton was also back in the city and their mutual interest in chemistry would have brought them together. They became close friends and collaborators for the rest of their lives.

It was an attraction of opposites. According to Hutton's first biographer and friend, both men had a passion for science and shared many opinions, yet

> Ardour, and even enthusiasm, in the pursuit of science, great rapidity of thought, and much animation, distinguished Dr Hutton on all occasions. Great caution in his reasonings, and a coolness of head that even approached to indifference, were characteristic of Dr Black. On attending to their conversation, and the way in which they treated any question of science or philosophy, one would say that Dr Black dreaded nothing so much as error, and that Dr Hutton dreaded nothing so much as ignorance; that the one was always afraid of going beyond the truth, and the other of not reaching it.[3]

Black's cool head and cautious reasoning was the perfect foil for Hutton's endless curiosity and 'powerful and imperious impulses'. It was not only in science that they differed. According to another biographer 'Dr Black was correct, respecting at all times the prejudices and fashions of the world; Dr Hutton was more careless, and was often found in direct collision with both.'[4] Despite their differences, Black and Hutton became 'inseparable cronies', meeting practically daily and often sharing a meal together. Hutton valued Black's opinions and tested his ideas on his friend – particularly over many years his developing theory of the origin of rocks and their destruction by erosion. Those who knew them recognised the genius of both, but to casual observers they could appear as eccentrics. After a discussion of food, they wondered why shellfish were considered delicacies, but land snails were not. As practical scientists they decided to experiment:

> having collected a number of snails, they had them cooked, and sat down to the banquet. Each began to eat very gingerly; neither liked to confess his true feelings to the other. Dr Black

at length broke the ice, but in a delicate manner, as if to sound the opinion of his messmate. 'Doctor,' he said, in his precise and quiet manner, 'Doctor, do you not think that they taste a little, a very little, queer?' 'Queer? Damned queer! tak them awa', tak them awa'!' vociferated Dr Hutton, starting up from the table, and giving vent to his feelings of abhorrence.[5]

Although Black was serious and reserved, he enjoyed Hutton's playfulness and humour and indulged him. After his discovery of carbon dioxide, Black's reputation as a chemist was international. Hydrogen, then called 'inflammable air', had been discovered by the English scientist Robert Boyle in 1671 and Black had demonstrated it to his students by burning pure hydrogen in his lectures, or by mixing it with air and exploding it. In 1766 he read a paper by Henry Cavendish which showed that hydrogen was much lighter than air. To demonstrate this Black had an allantois (a calf's birth sack) filled with the gas and, rather than demonstrate it to his students, invited a group of friends – including John and George Clerk – to supper in his house. We can guess that Hutton was the inspiration for the stunt which followed. When the bladder was released, it floated upwards and appeared stuck to the ceiling. Hutton invited the guests to guess how this had been achieved and the consensus opinion was that there was a thread attached to the top of the bladder, which went through a hole in the ceiling and was pulled upwards from the room above. It was only when the bladder was brought down and no thread was discovered that they accepted Black's explanation of the cause.[6]

Neither Black nor Hutton grasped the commercial implications of the discovery. Hydrogen balloons were developed in Paris and demonstrated in Edinburgh in 1784, when James Tytler became the first person in Britain to ascend in a balloon. Tytler's venture was expensive, but succeeded after several attempts on 25 August 1784, when his balloon rose a few feet from the ground. Two days later he managed to reach a height of 350 feet, travelling for half a mile between what is now known as Holyrood Park to the village of Restalrig. Later trials were less fortunate. In October his balloon only took off after Tytler got out of the basket, to the disappointment of the crowd. Having previously been 'the toast of Edinburgh', he was ridiculed and called a coward. His last flight was on 26 July 1785. Tytler's fame was eclipsed by the Italian diplomat Vincenzo Lunardi – the self-styled 'Daredevil Aeronaut' – who carried out five sensational flights in Scotland, including an ascent from the green at Heriot's Hospital, watched by a crowd of 80,000. He inspired a ballooning fad and ladies' fashions in balloon-shaped skirts and hats. The 'Lunardi bonnet' is mentioned in the poem 'To a Louse' by Robert Burns.[7]

★ ★ ★

While Hutton and Davie were establishing their chemical works, a fierce controversy was raging in newspapers, magazines, coffee houses and in town meetings over proposals to cut a canal between the rivers Forth and Clyde – creating a quick and cheap means of moving goods and people between east and west Scotland. Canals were not new; the French had opened the enormous Canal Royal de Languedoc* in 1681, the Bridgewater Canal in the north of England had been completed in 1761 and was extended five years later. A central Scotland canal had been suggested as far back as the reign of Charles II and rival routes had been surveyed many times, but the conclusion was always that the cost was unaffordable. What made this time different was the new spirit of enterprise and innovation that was sweeping Scotland and the growing prosperity and commerce, which provided both an economic incentive to construct a canal and the wealth to finance it.

Glasgow tobacco merchants who had built considerable fortunes by trading with the English colonies in America financed a parliamentary bill which would allow them to raise funds and construct a four-foot-deep canal. Their motive was to make it faster, safer and cheaper to re-export tobacco to the large markets of Europe than sending ships around the perilous waters of Cape Wrath and the Pentland Firth. A shallow canal would only be able to take barges, but they argued that it would cost less and be quicker to build than a deeper canal and as the Clyde in Glasgow was only four feet deep there was not much point in going any deeper. They had support from harbour owners, particularly those at Bo'ness on the Forth,

* Now called the Canal du Midi.

who feared loss of their port duties if sea-going ships were able to pass directly from one side of the country to the other. The opposing view was taken by the merchants of Edinburgh and the landowners of the east of Scotland. They proposed a seven-foot-deep canal to accommodate vessels of 40 tons (later enlarged to take ships of up to 60 tons). Hutton was gripped by this argument and wrote a number of articles in his robust style – 'in which the grave and the ludicrous were occasionally employed,' says Playfair[8] – but seems to have been more concerned over the route and the physical obstacles to be overcome than the commercial objectives.

The project was the largest financial commitment in Scotland since the failure of the Darien scheme at the beginning of the century, but it was also an immense engineering challenge. The new waterway was to be over 35 miles (56 km) long, connecting the River Forth near the mouth of the River Carron to the Clyde at Bowling, West Dunbartonshire. It had to be cut through 20 different rock formations and a soft bog. It needed 20 locks on the eastern section to lift vessels 150 feet (45 metres), and 19 on the western section to take them down again. A reservoir had to be built and water from five lochs used to keep it filled. In 1767 a compromise was reached between the two sides with the publication of a report by the engineer John Smeaton for the Board of Trustees for Fisheries, Manufactures and Improvements, which concluded that a deeper canal would be only marginally more expensive and had a greater chance of becoming profitable.[9]

A new parliamentary bill gained Royal Assent in 1768 and an initial fundraising brought pledges from the great and good of Scotland, including the Dukes of Buccleuch and Queensferry, seven earls, a dozen baronets, three knights, lairds, merchants, and the lord provost of Edinburgh and provost of Glasgow on behalf of their cities. There were also English investors, led by the Duke of Bedford. The total of pledges was £150,000 – the estimated cost of the construction – but in the first two calls for cash subscribers had only to pay ten per cent of the value of their shares. The largest shareholder was the self-made merchant and politician Sir Lawrence Dundas, whose estates straddled the proposed route and could therefore benefit by using the canal to transport his coal, grain and other produce. Two of Hutton's university friends subscribed for shares – Sir John Hall for £1,000 and John Clerk for £500, although Clerk appears to have held his on behalf of the Carron Ironworks, where presumably he was a shareholder.

Neither Hutton's name, nor that of his close friend George Clerk (by this time calling himself George Clerk Maxwell) are mentioned in the parliamentary bill as initial shareholders in the Forth & Clyde Navigation Company, but both joined the committee of management, which was to oversee the construction of the canal. It was set up even before the Act of Parliament was obtained and it met in Edinburgh under the chairmanship of the lord provost. Clerk Maxwell is listed as a member from the first meeting, and it is likely that he suggested Hutton be approached to join to advise on the geological issues which would be confronted during the digging. He had previously commissioned Hutton to undertake geological surveys on his own land and had clearly been impressed by his knowledge during their tour of the forfeited estates three years earlier. By the fifth meeting in December 1767, Hutton had joined the group and probably became a shareholder.[10]

Decision-making in the company was split and confused from the start. The governor and council – equivalent to the chair and board of directors of a modern company – met in London, close to Westminster. The project would need political support and several more Acts of Parliament to extend the powers of the company before the canal was completed. The Edinburgh committee was a much more fluid and informal group. Initially 20 people attended, including Sir John Hall and his brother William, although they seem to have dropped out early in the history of the project. At a meeting on 15 April 1768, a committee of management was elected, including George Clerk Maxwell and Hutton, which effectively set the construction in motion by appointing managers, engineers and legal agents and drawing up detailed job descriptions and lines of responsibility. It also set freight charges, trying to strike a balance between rates cheap enough to attract business, but which would make the project economically viable. Members of the group were unpaid and received no expenses, even when they made site visits. As the geological, engineering and financial problems of the project unfolded, Hutton became deeply involved in the work.

The enormity of the task soon became apparent. The canal had to be cut through ground which included areas of hard whinstone, which had to be blasted with gunpowder and made progress very slow. At the other extreme, the Dullatur Bog presented unforeseen challenges. Various routes had been examined to find ways around the area, but all proved to be unsuitable. Turning it into a reservoir was also considered but abandoned as impractical. There was no alternative to going straight through. A team of 50 workmen was employed exclusively rebuilding the banks as they sank into the mud – it took 55 feet of earth and stone before the sides of the cut were stable. Even then, the bog threw up unexpected challenges as bodies began to be exhumed. Most were the corpses of Covenanters fleeing Montrose's army after the Battle of Kilsyth in 1645.[11] According to one report:

> a number of swords, pistols, and other weapons were dug out; also the bodies of men and horses, and what seems somewhat marvellous, a trooper, completely armed, and seated on his horse, in the exact posture in which he had perished.[12]

The construction did bring some benefits to the local communities. According to a 19th-century account, wheelbarrows had been unknown in Scotland until imported from England for the construction. There was also employment: a man and a horse were paid only a shilling (a twentieth of a pound) a day, but labour was so scarce that women also found employment. Jenny Bull managed to build herself a house in Kirkintilloch High Street with her earnings – although she said she had worked 'like an Ox'.[13] Getting enough manpower was, however, a constant problem. Locally hired workmen tended to desert the site and go back to the land at harvest time, so the company imported 'navvies' (men working on inland navigation) from Ireland, the Highlands of Scotland and England. This was not a complete solution – they were too far from their homes to return at harvest time, but many left to find casual farm work during the summer to give themselves a break from the monotonous digging.

The supply of labour was not the only personnel problem the Edinburgh committee had to deal with – there were also management disputes. John Smeaton, the chief engineer, was frequently absent on other projects and left more and more of the task of over-seeing the construction to Robert Mackell. A Glasgow engineer who had been engaged on dredging the Clyde in Glasgow, he had twice surveyed the route of the canal (the second time in 1764 with James Watt, who had previously been a scientific instrument maker for Joseph Black at Glasgow University). He favoured a different route for the canal but lost out to Smeaton. He became involved in frequent arguments – with the Carron Ironworks over the supply of water and with James Watt, who had gone on to become chief engin-eer of the Monkland Canal. Mackell accused Watt of poaching his workforce, but when the Edinburgh committee investigated it was shown that both companies were offering the same daily rates.

The committee was involved in an astonishing amount of detail. It had the responsibility of resolving conflicts with landowners and with its own shareholders. The Carron company proved particu-larly difficult, demanding a spur channel be cut to give direct access to its factory and withholding payment on its shares to apply pres-sure on the canal company. Years of negotiation and threats of legal action were needed to settle the argument.

Committee members made frequent visits to the construction site, discussing gradients, expenses and water supplies with engin-eers and surveyors. They debated the design of aqueducts and locks, negotiated contracts for the purchase of tools and materials, such as wood, lime, stone and 'Pozzolana earth' – a volcanic deposit, which sets hard under water and could be used as a cheaper substi-tute for cement. They sanctioned contracts with 'undertakers' – not funeral directors, although they doubtless had to do that too since many were killed in the construction – but subcontractors who hired the labour force. They arranged for medical treatment for injured workmen, first by paying local doctors then, when the cost became too high, having incapacitated men transported to Glasgow to be treated in the infirmary. They even made an order that all tools should be labelled 'FCN' as a precaution against theft.[14]

Meetings took place in the Exchange Coffee House in the High Street of Edinburgh's Old Town, but the meetings often went on so long that the committee adjourned to nearby Fortune's Tavern for evening refreshment. Hutton and Clerk Maxwell were more con-scientious attenders than any of their colleagues, Hutton being present at 52 of the 78 meetings held between 1767 and 1774 and Clerk Maxwell at 54. In addition they travelled to site meetings – often being away for two or three days at a time – which other committee members avoided. Hutton made at least nine of these trips and would have both used his expertise and gained it in discussing the different terrain which lay in the canal's path. The minutes of the site meetings give an indication of the amount of work the unpaid members undertook.

In June 1768 Hutton, Clerk Maxwell and two colleagues met Smeaton at Kerse, near Grangemouth, to decide between two possible routes for the canal's entry into the River Carron, and after walking over the ground staked out by Smeaton, advised in favour of the more westerly route.[15] In April the following year four committee members, including both Hutton and Clerk Maxwell, met at Falkirk with Mackell, the surveyor John Laurie, and the clerk of works, Alexander Stephen. The meeting lasted for three days, when, despite 'very bad rainy weather', they inspected all the work so far accomplished and a quarry at Glenfuir, just outside Falkirk. They then debated how a temporary cut could be made to transport the stone from Glenfuir, and visited a plot of land they had bought to store timber, lime and other materials.[16]

Five months later they were back in Falkirk again and inspected the whole of the line of the canal together with 'locks, aqueducts, bridges, and quarries'. Although they 'had the satisfaction to find everything going on very well,' they left behind them a stream of instructions and exhortations. Stephen was asked to adopt a new system of accounting, while Mackell was instructed not to build any more aqueducts until the committee had approved their design. Finally, they 'ordered Mr Mackell to make enquiry as to what quarries could be got at the western end of the canal and to consider the propriety of opening these quarries and making roads to them and what may be the depth of tirring [overburden] and to report' and 'ordered Mr Mackell to give proper directions for carrying on those quarries now working through the winter and to prepare and lay a state of them before the committee'.[17]

Hutton was also given some special tasks. For example, he was asked by the committee to investigate the cheapest way to buy timber and to use his medical experience to give an option on a doctor's bill.

He took less interest in the financial side of the project, being asked only once to approve the accounts, but attended 23 of the general, special and quarterly meetings of the canal shareholders.

It is clear that Hutton had a passionate interest in seeing the canal completed and in its financial success and devoted a lot of time and energy to it. In 1777 he published his first public paper, *Considerations on the Nature, Quality and Distinctions of Coal and Culm*, which argued that culm – soft 'small coal' produced in central Scotland and used largely in the manufacture of bricks and lime – should be taxed less than coal. Since Hutton had no interests in coal- or culm-mining the pamphlet seemed to be designed to make transport of culm by canal more attractive. It was part of a political campaign mounted by the canal company, which was ultimately successful, the tax being reduced in December 1777 and the volume of culm carried on the canal rising thereafter. But despite this small success the canal project was dogged by financial problems from the outset.

The initial estimate of the cost of the canal was hopelessly optimistic. Geological problems and management errors meant that the project was behind schedule and over budget from the start and the company had constantly to try to raise more money. It had been intended that the canal could be in use and carrying cargo to the port at Grangemouth as soon as it reached Kirkintilloch and that this would provide a source of revenue to finance the continued building, but it took five years to reach that stage. The enthusiasm of the initial shareholders when they registered to buy shares was not sustained when the company tried to call in the funds – many were slow to part with cash, or defaulted, leading to threats and even litigation. The situation was made worse by the financial crash of 1772, which led to the collapse of the Bank of Ayr and the bankruptcy of dozens of its shareholders, many of whom had also pledged to buy shares in the canal company. Hutton, however, was never among the defaulters, nor did he join the petition to persuade the government to rescue the project with an injection of funds.

As the canal approached its western end the Glasgow merchants became anxious that their interests were being ignored by the Edinburgh committee. They set up their own management board, began direct negotiations with the council in London and slowly cut the Edinburgh committee out of the oversight of the project.

Hutton's direct involvement came to an end at this stage, although he continued to attend shareholders' meetings. It is doubtful that he saw any return on his investment. The canal took 22 years and required a government subsidy to complete. All that time the share price remained static and the company, which was heavily indebted, was not able to pay a dividend.

Chapter 9

Wrote by the finger of God

HUTTON'S RETURN TO Edinburgh from his farm in Berwickshire meant again sharing a home with his three spinster sisters, Isabella, Sarah and Jean. We don't know precisely where they were living, but it was possibly still in the tenement in the narrow Gosford Close, which had been his father and mother's family home. This was a problem for James because he had to share his bedroom with his books, scientific equipment and his growing collection of rocks. He complained in a letter:

> I am obliged to make one chamber serve me for laboratory, library and repository for self and minerals, of which I am grown so avaricious, my friends allege that I shall soon gather as many stones as will build me a house; the truth is, build I must or be separated from my studies; my ambition is to make a spacious library, where the books shall consist, in the ancient manner, of tables of stone, and (without any mystical sense) wrote by the finger of God alone.[1]

In 1770 Hutton used the affluence that the partnership with Davie had brought him to build a house for himself and his sisters. Edinburgh was expanding rapidly, with airy new squares and wide streets to the south and a new settlement planned and being built across the recently bridged Nor' Loch to the north. Prosperous merchants and professionals were moving out of the cramped and insanitary Old Town and building themselves larger houses, facing gardens. One of the early movers was David Hume, who left the High Street for South St David's Street in the New Town. Hutton declined to move out of the Old Town and bought a large plot of land in St John's Hill, a short street which ran from the Canongate down to the foot of Arthur's Seat, the extinct volcano close to Holyrood Palace.

The area took its name from the Knights of St John of Jerusalem, who had owned the land in the 15th century. It was undeveloped until 1766 when John, second Earl of Hopetoun, built the property now known as St John's Land and a row of tenements facing the Canongate. Access to St John's Hill was through a wide pend, which still exists. The Order of St John still maintains a house in the street, which was already standing when Hutton moved there.[2] The building he commissioned was not in the fashionable Georgian style of George Square or St Andrew Square, but a detached two-storey stone building, more resembling a Scottish farmhouse than the town house of a successful businessman. It was described by a later descendant as 'a rather quaint and grim specimen of Scottish domestic architecture'. It stood back from the road, was overshadowed by trees, and reached through a gateway and a short walk. 'On the right, on entering was a long dining room . . . on the left a small parlour and upstairs a long drawing room and a number of bedrooms all stiffly furnished and with four-poster beds.'[3]

At the back was a green and a garden, in which Hutton continued to grow plants which might yield industrial dyes or chemicals or demonstrate possible improvements in agricultural productivity. The property was big enough for Hutton to sell part of the land to John Robison, who became professor of natural philosophy at Edinburgh University. He built a house for his family there. Hutton's new home had reception rooms large enough to host his friends over dinner discussions, bedrooms for himself and his sisters, and a laboratory, where he could continue to conduct chemistry experiments and space to store his growing collection of fossils and stones.

The medical student Silas Neville visited him there in 1772 and wrote back to Dr John Manning, who had introduced him to Hutton: 'In consequence of your kind recommendations I was received with much civility by Dr Monro [Professor of Anatomy] and Dr Hutton, particularly the latter, who has been of signal service to me. He is the oddity you described, but at the same time a mighty good sort of man. His study is so full of fossils and chemical apparatus of various kinds that there is hardly room to sit down.'[4]

Neville described himself as 'a dissenter of liberal sentiments and good fortune', and on occasion was shocked at Hutton's views:

> Dennison [a fellow student] and I dined at Dr Hutton's by invitation. He shewed us a piece of limestone rock of Gilbralta [sic] which is everywhere full of bones and parts of bones. Dr Hutton is for having all laws against bribery & corruption abolished and everyman sell his vote as he does anything else. I knew that political principles here were none of the purest but did not imagine that they proceeded to this length. Dr Black, too, who dined with us joined Hutton in thinking that we enjoy perfect liberty and have no political evils to complain of. Gracious God, that men can be so blind! Was uneasy on being engaged to this visit, but (thank God!) no disagreeable circumstances arose from it.[5]

On another occasion, Hutton again surprised Neville, although this time with his scientific rather than his political views. 'Had a visit of Hutton. He is a most enthusiastic chemist. He actually acknowledges having tried to discover the philosopher's stone.'[6] Whether Hutton genuinely believed either of these statements, or said them merely to provoke an interesting argument, is open to question. His surviving papers and letters hardly mention politics and he was not the only scientist not to dismiss alchemy out of hand. Neville appears to have been a rather spoilt and arrogant young man ('at last [I] tired of the insipid life of a gentleman without employ, I determined to study physic') and Hutton, for all his kindness, may have been having fun at his expense, with Black playing along.

He would not have got away with such remarks during meetings of the Philosophical Society. By the time Hutton was elected to the society in 1768, possibly introduced by his friend George Clerk Maxwell who was one of its joint secretaries, it had established itself as Edinburgh's leading forum for the introduction and criticism of ideas. Its president was Lord Kames who, although 72 at the time of his election, was still a formidable intellect and activist. He had been a senior judge and written books on the law and philosophy, and published volumes of criticism. He was an improving farmer and through his membership of the Board of Trustees for Arts and Manufactures, the Commission for the Forfeited Annexed Estates, and the British Linen Company, had a strong influence over economic policy.

Other members included William Robertson, principal of Edinburgh University, David Hume and Adam Ferguson, professor of moral philosophy at the university. There was strong representation from the medical profession and from landowners, who had interests in agriculture, mining and manufacturing. The society was well connected with similar intellectual societies throughout Britain – particularly with the Royal Society in London – and had members in Pennsylvania, South Carolina, Jamaica and other West Indian islands, and contacts in Europe, Russia and as far away as China and Siberia.[7] At meetings members read essays, which were then subjected to criticism and debate. In the seven years up to 1771 it published three volumes of papers.

Hutton was now in his forties and until this point could be considered a dilettante. He was not rich but had used the comfortable income his father had left him to dabble in areas that took his interest for a period. He had been a trainee lawyer, a medical student, a farmer, a scientist, a businessman. He was known as a wit and an amusing dinner companion who was interested in everything – but unlike his close friend Black he had made no major discovery, and unlike Watt he had invented nothing of consequence. He did not teach at the university or give public lectures and he had published nothing since his medical dissertation. His letters betray no major ideas or coherent new theories; they do not hint at his political views or give any commentary on the major news of the day, although written in periods when Britain was frequently at war. The Edinburgh Enlightenment was reaching its peak – with major advances in medicine, mathematics, science, philosophy and law – but Hutton had little to say. Some of its leading lights – like the historian William Robertson, the doctor William Cullen, Joseph Black in chemistry, David Hume in history and philosophy and Adam Smith in moral philosophy (although he was at this time a commuter from Glasgow) – had national and international reputations. Hutton was a minor character on the fringe of its debates. That was about to change.

Playfair claims that Hutton addressed meetings of the Philosophical Society and exposed his argument to critical analysis,[8] although nothing by him appears in the published accounts of the meetings. The only contribution we do know about was a short description of curious stripes of dead grass observed on Arthur's Seat, but the account of his talk was not published until many years later.[9]

An indication of the profound change which was taking place in his thinking and his writing came with his first publication, the pamphlet on the difference between culm and coal (see Ch. 8). It announced itself as 'inquiries, philosophical and political, into the present state of the Laws and the questions now in agitation relative to the taxation upon these commodities', and was signed by 'Doctor James Hutton, physician, Edinburgh, in a letter to a friend'. Although only 35 pages long, it was a well-argued and fluently written case for changing the law setting the excise duty paid on transporting each commodity.[10] Culm attracted a lower duty than coal, but the problem was that there was no legal definition of culm. Excise officers could easily decide it was coal and thus subject to a higher rate of tax. Hutton clearly had a vested interest – he was a shareholder and a member of the management of the Forth & Clyde Canal Company; a reduction in the duty on cargo carried along its waterway would lead to higher volumes and thus greater revenue for the company. He may have been encouraged to write it by George Clerk Maxwell, also a shareholder, and by Lord Kames, who had a strong interest in helping the economic development of Scotland.

Against the scale of the canal project, the change the pamphlet hoped to bring about was a fairly minor contribution, but it is the style in which it is written that makes it so interesting. In contrast to his letters, where thoughts and observations are jumbled together without clear punctuation or a narrative thread, the pamphlet is written in short paragraphs which flow naturally from one to the next to form a plausible case for making a distinction between coal and culm – although they might look the same and if burnt in a domestic hearth produce the same heat. It makes not only a scientific case for distinguishing between the two substances – and proposes a simple practical test – but argues that it is in the economic interests of the country for the change to be made. Up until this point Hutton had not demonstrated any interest in economics or industrial policy. Either he had learned how to present an argument from listening to his fellow society members and from reading their papers, or he had a helpful and thorough editor – Joseph Black perhaps?

Interestingly the pamphlet provoked a response in the sort of intemperate language that Hutton may at one time have used, but he had restrained himself on this occasion. Where Hutton put his name to his case, the counter-blast is signed 'A friend to the Revenue', and presented as a letter to the Commissioners of Customs:

> Having lately met with a little pamphlet, wherein you are treated very illiberally, for daring in the course of your duty to differ in opinion from some interested people in Scotland; I cannot refrain from offering you a few Remarks on this wild futile rhapsody, written by an author who is deficient in argument as he is in good manners.[11]

Whereas Hutton's text is coolly and logically argued, the response accuses him of 'licentiousness' and 'disclaiming the aid of chemistry and natural history to which all rational people have recourse in a question of mineralogy' and quoting no authority but his own '*ipse dixit*'. Hutton, however, was the eventual winner. His pamphlet was accompanied by a ten-month long lobbying campaign, which was ultimately successful in getting the law changed.[12]

★ ★ ★

Hutton's intellectual circle was growing far outside the Edinburgh scientific and literary establishment. As his theories developed in his mind, he became hungry for evidence and information. He read avidly and wrote to naturalists and scientists offering to exchange research findings and samples. He had also absorbed an extensive knowledge of British geology and geography, learned from the journeys around the east of England he had made in 1752, his years of farming in the Scottish Borders and his trip to the north of Scotland with George Clerk Maxwell. The detailed mental map he carried in his mind was demonstrated in an exchange of letters with Sir John Strange, a fellow of the Royal Society in London who had published a number of scientific papers. They were written between 1769 and 1770, when Strange was the British Resident (diplomatic representative) in Venice, then a separate state.

Hutton was introduced to Strange by other amateur researchers, the lawyer Daines Barrington and the Welsh naturalist Thomas Pennant, who had probably met Hutton during a tour of Scotland in 1769.[13] Letters from Barrington, Pennant and Strange to Hutton have disappeared with the rest of his papers, but some of his replies to them survive. From them it appears that Barrington had missed seeing Hutton, but had asked him to identify some of the fossilised creatures in Strange's collection. Hutton's reply makes it clear that he had another purpose in collecting his samples. He was known as a fossil collector, but in reality he was a rock collector with a different purpose in mind:

> As I never made shells my study I cannot tell the names of the different fossil shells found in Scotland; I have always been in use to look upon them as marine bodies when I meet with them in minerals and considered them in that general light without examining into their particular kinds which is the thing perhaps that Mr Strange wants to be informed about.
>
> The limestone strata in Scotland, like those in England, for the most part, contain shells & madrepores [corals], tho I don't remember of seeing any great variety of their kinds; this however I paid little attention to, as the stone was always the object of it and not the particular figure.

Despite their differing interests, Strange and Hutton kept up a detailed correspondence: four of Hutton's letters survive and there were presumably a similar number from Strange. In them Hutton draws on his first-hand surveys of eastern and southern England, the Scottish Borders and the north of Scotland. Although his letters at the time of those trips might suggest a casual interest, it is clear he was amassing information in a more detailed and systematic way. For those areas he had not been able to visit, such as Ireland or the Western Isles of Scotland, he relied on reports from others, samples collected for him or perhaps purchased. For example, he writes: 'I have a piece of a large belemnite* found in clay in the Isle of Sky.' Although not able to identify particular fossils, he is able to

---

* Squid-like creatures from the Cretaceous Period.

give Strange detailed information about the areas and the type of rock in which fossils might be found, sending him a map marked up with 'a general sketch of the mineral geography'.[14] He despatched a packing case of fossils and commissioned others to find samples that would interest Strange. We don't know what Strange did in return, but it was not a one-way transaction. Hutton wrote:

> I think myself greatly obliged to our worthy friend [Pennant], for this agreeable correspondence and opportunity of trading, so much to my advantage. Miser as I am of my treasures, I must open my purse with pleasure in expectation of so valuable a return, at the same time I must express my private sentiments of this traffick; that whatever may be my expectations from Mr Strange's equity as a merch[t], my dependence must be more upon his generosity as a Philosopher.[15]

Although it would be more than a decade before Hutton's completed theory was ready to be presented to the public, the letters make clear that his interest in fossils was as evidence of the sedimentary formation of rocks and it had not gone unremarked that most fossils were found above sea level:

> We are both engaged in the same pursuit; the field is wide, we cannot therefore be supposed to keep the same track, altho' we may arrive nearly at the same end. My attention has been chiefly upon the various substances that enter into the composition of the mineral kingdom in general; and being neither botanist nor zoologist in particular, I never considered the different kinds of figured bodies, found in strata, further than to distinguish betwixt animal & vegetable, sea & land objects; the mineralisation of those objects being more the subject of my pursuit, than the arrangement of them into their classes & I cannot therefore pretend to give any information about particular kinds of which I do not even know the names.[16]

As far as we know, after his trip to the Low Countries 20 years previously, Hutton never again left Britain. But he was interested in information from other parts of the world and valued exploration and the experience of travellers. He made contact with Joseph Banks shortly after the explorer's return from the three-year voyage of Captain Cook in July 1771. Cook's ship, HMS *Endeavour*, had visited South America, Tahiti, New Zealand, Australia, Indonesia and South Africa, map-making, recording the flora and fauna and collecting botanical samples. Banks had intended to join Cook's second voyage the following year, taking with him the Swedish naturalist Daniel Solander and the young Scot James Lind, but a disagreement with the Admiralty led to him being excluded. Instead the trio travelled by ship to Iceland, visiting the Hebrides on the way and making the first scientific survey of Fingal's Cave, on the isle of Staffa, with its distinctive hexagonal basalt columns. On their return they visited Edinburgh and met Hutton.

Hutton must already have known Lind by the time of their meeting with Banks and Solander. Lind was born and brought up in Edinburgh and had studied medicine at the university. His uncle (also James Lind) was a prominent physician with a house in Princes Street. Although only 36 at the time of their meeting (Hutton was ten years older), the younger Lind was already known as a physician and scientist. His medical thesis had been on the outbreak of fever in Bengal in 1762, which he had seen as a doctor in India, and he had published papers for the Royal Society on his observations of the transit of Venus across the sun and an eclipse of the moon. In early 1772, when Lind was still expecting to accompany Banks on Cook's second voyage, Hutton wrote to him at Banks' London home with advice about the readings he ought to take on the journey.

In contrast to his letters to George Clerk Maxwell and James Watt (see next chapter), this message is straightforward and business-like, with none of the vulgar asides that peppered his other correspondence. This is Hutton the serious scientist, who advises the younger man to 'begin with method and pursue . . . with unwearied steadiness . . . astronomical and meteorological measurements and observations of the sea – its colour, depth, temperature streams and tides regularly recorded; and whatever may be occasionally observed on or in it.' On land, Hutton has another request:

You must consider that to us the curious observations of this country are very interesting, but it is the more common observations of a distant country that contribute to our instruction. Therefore, never lose sight of the common observations to be made on such a country, which will afterwards, if full and accurate, come to be extremely useful in being compared with those of this or any other country. Is not this truly curious? – natural history in perfection.[17]

He asks for observations and soil and rock samples and adds, 'NB a bag of gravel is a history to me and . . . will tell wonderous tales.' There are instructions about how to gather gravel samples and a request for descriptions of the landscape from whence they came, including the height of mountains. Helpfully he gives Lind three different ways of measuring the altitude of peaks. Intriguingly he encloses two receipts for 'diving machines' which he had sent by cart. 'Pray exert yourself to get the machine down to the greatest depth; let no cord be spared, take enough to reach the bottom of hell and get us samples up from Tartarus.'* He ends with good wishes for a safe and pleasant trip, but adds his own wish for rock samples: 'When at the social board forget not the dainty tables of stone; mind every one of them is wrote upon by Gods [sic] own finger. I have known a duodecimo [the size of a small book] of this kind do more credit to the author than a quarto bible.'

On the way home from their trip to the Hebrides and Iceland, Banks, Lind and Solander visited Hutton and Black in Edinburgh and remained in the city for three weeks. Banks brought presents of Icelandic 'petrified vegetables and incrustations, formed from the water of boiling springs'. Given his later interest in granite, Hutton would have found Banks' description of Fingal's Cave fascinating. Banks wrote a detailed account of his survey, which was published by Pennant in his book *Tour in Scotland, and Voyage to the Hebrides*. Banks had also brought back large quantities of Icelandic lava, carried as ballast in his ship, the *Sir Lawrence*. Samples are still to be seen in London at Kew and the Chelsea Physic Garden.[18] It would be surprising if Hutton had not also begged a piece.

---

* The deep abyss in Greek mythology.

Hutton and Banks obviously got on well and Hutton's subsequent letter to Banks in London is in a more intimate style than the one to Lind. He encloses receipts for boxes of books and stones which he had sent as gifts. The main part of the message consists of Hutton laying down a marker for any future exploration which Banks might make – not to be included himself, but to benefit from the observations and findings which might result. He suggests some regions have already been studied, and then ventures a joke at Banks' expense: 'in new countries, on the contrary you neither know where to go and if you should there is nothing to be seen at least little of the *Solum Sine veste* [without clothes] and Dame Natures [sic] petticoat is not so easily lifted as that of Princess Obrea' – an allusion to the Tahitian woman with whom Banks was alleged to have had a romantic liaison.[19]

Hutton was not content just to collect information from others. Although now resident in the city, he continued the measurements and experiments he had begun while farming. He recorded daily temperature readings to inform his ideas about the effect of heat on growing barley and constructed tables of the 'vegetating heat' during the growing season. A cool summer would delay the harvest and in very poor years might mean the crop was worthless. He estimated the difference the angle of the sun would make to growing conditions in other locations and measured the temperature difference at the base and the summit of Arthur's Seat. His estimates of the temperature difference at different latitudes and altitudes correspond remarkably well with modern calculations.[20]

Chapter 10

Lord pity the arse that's clagged to a head

IT IS POSSIBLE that Hutton met James Watt through his work on the Forth & Clyde Canal. Watt had been involved in the project even before Hutton, surveying possible routes with Robert Mackell in 1766 on behalf of the Glasgow merchants and visiting London to help with the parliamentary lobbying to get an enabling Act. He was not impressed, writing to his wife: 'I think I shall not long to have anything to do with the House of Commons again – I never saw so many wrong-headed people on all sides gathered together.'[1]

In 1770 Watt began work on the Monkland Canal, a short 12-mile-long channel designed to bring coal from the mines to the east of Glasgow into the city, but it was a project fraught with problems and as chief engineer it fell to him to resolve many of them. There was conflict with the Forth & Clyde Canal Company over access to water and with his former colleague Mackell over the supply of labour. Finding and managing contractors who were capable of doing the work and could guarantee a workforce was a nightmare for him. He had personally to survey, level, plan and stake out the ground and then measure the volume of earth excavated on which the 'undertakers' (contractors) were paid; he had to negotiate with them, supervise their work, check their accounts and pay them. It was a job he hated: he wrote to a friend, 'I have been cheated by undertakers and clerks and am unlucky enough to know it. The work done is slovenly, our workmen are bad and I am not sufficiently strict . . . I would rather face a loaded cannon than settle an account or make a bargain.' He recognised he would never make a businessman, adding: 'I find myself out of my sphere when I have anything to do with mankind. It is enough for an engineer to force nature and to bear the vexation of her getting the better of him.'[2]

The project was short of finance from the start, with the familiar problems of shareholders failing to pay up once their initial enthusiasm had waned. Watt's salary of £200 a year was frequently in arrears, and on one occasion he had to raise money from the Glasgow Thistle Bank on his own account to pay contractors. Eventually the company ran out of cash and in the aftermath of the 1772 financial crisis the proprietors reasoned that it would be impossible to get more from the shareholders. Watt left in July 1773 after making a last inspection of the seven miles he had completed. He was employed for a while on other surveying work – for the Crinan Canal and for a Highland canal (eventually built many years later as the Caledonian Canal), but his heart was not in civil engineering and he took the jobs merely to earn a living.

Hutton and Watt had a mutual friend in Joseph Black. While Black had been a professor at Glasgow University, Watt had been an instrument maker and had designed and constructed many pieces of apparatus for him. Black took to the young man, not only admiring his mechanical ability, but as a person 'remarkable for the goodness of his heart and the candour and simplicity of his mind, as for the acuteness of his genius and understanding'. Watt did not have a university education, but nevertheless had wide knowledge and was renowned for his problem-solving ability. Another professor, John Robison, who was also at Glasgow (and later to join Black at Edinburgh University) described meeting him: 'I saw a workman ... but was surprised to find a philosopher as young as myself and always ready to instruct me. Everything became to him a subject of new and serious study. Everything became science in his hands.'[3]

Hutton appears to have met Watt sometime in 1772 or 1773, possibly introduced by Black. Although ten years younger than himself, Hutton saw in his new friend a kindred spirit – relentlessly curious about the world and incapable of seeing a problem without trying to devise a solution. Both men defied pigeon-holing. Watt was no more a mere engineer than Hutton was just a doctor, or a farmer, or a chemist. Their relationship – although often separated by hundreds of miles – was to remain a close one for the rest of their lives. According to Playfair's biography, Hutton 'had an uncommon facility in comprehending mechanical contrivances; and, for one who was not a practical engineer, could form, beforehand, a very sound judgement concerning their effects . . . he would rejoice over Watt's improvements to the steam engine, with all the warmth of a man who was to share in the honour and profit about to accrue to them.'[4]

Watt had become interested in steam power while at Glasgow and began to develop his ideas when he was asked to repair a model of a Newcomen engine at the university. In doing so he realised that the design could never be efficient because it required the cylinder to be heated and cooled at every stroke. He built a larger model on which he demonstrated a major innovation – a separate condenser. In 1765 he was asked to install an engine for Dr John Roebuck, whose acid factory in Prestonpans had supplied Hutton and Davie with glass vessels for their sal ammoniac production. Roebuck had become a partner in the Carron Ironworks at Falkirk and had leased coal mines in West Lothian from the Duke of Hamilton. He asked Watt to build him a steam engine to pump water from his mine and, in return for paying off the £1,000 of debt Watt had incurred develop- ing his prototype, took a part share in the patent. But the Carron Works was unable to machine the piston cylinder to an accurate enough specification and the engine never worked satisfactorily.

Roebuck, who had invested all of his own and his family's money in the iron and coal venture, was bankrupted in 1772. Watt turned first to Black, who lent him money and then to Matthew Boulton, owner of a Birmingham engineering company, to help him develop his machine. Watt, Roebuck and Boulton already knew each other, and Watt had visited Boulton's factory and was impressed by the range and quality of the work there. Boulton took over Roebuck's share of the patent and invited Watt to join him in the West Midlands to work on an efficient engine which could be manufactured in large numbers.

Watt's move from Scotland to England in May 1774 was to mark a decisive change in his life and his fortunes and he never returned to live in his native land. He intended the journey to be more than just a house-moving and he invited Hutton and Black to join him and to explore the country on the way. Black either could not or did not want to go, so Watt and Hutton set off together. The trip took two weeks and Watt wrote to his father that it had been a pleasant excursion during which 'nothing remarkable happened'. The expedition was much more eventful for Hutton, who was already turning over in his mind the theory of the development of the Earth he was to publish 14 years later and mentally collecting evidence. On the way they visited halite (rock salt) mines in Cheshire and marvelled at the concentric rings on the roof of the mine. Many years later Hutton was to cite this observation in his paper of the age of the Earth.[5]

Hutton made the most of his stay in Birmingham, which, like Edinburgh, was going through an explosive period of ideas and innovation. His gregarious nature meant that he quickly made new friends. Through Watt he met Boulton and through Boulton the businessmen, scientists, artists, writers and philosophers who were at the centre of the city's enlightenment and gathered every full moon at a dining club called the Lunar Circle.* Besides Boulton, the leading members included the potter and social reformer Josiah Wedgwood, the theologian and chemist Joseph Priestley, the chemist James Keir from Scotland, the physician and philosopher William Small, a fellow Scot, and Erasmus Darwin, grandfather of Charles Darwin.

Darwin had been educated at Cambridge, but then studied medicine at Edinburgh. He had set himself up as a doctor, first in Nottingham and then in Lichfield, where he established a successful practice. Like Hutton, Darwin had wide interests. He was a botanist, wrote poetry and was an inventor, devising a carriage steering mechanism, a manuscript copier and a speaking machine. Hutton was not in Edinburgh when Darwin undertook his medical training, so it is unlikely they already knew each other, but they quickly became friends and Hutton arranged to visit Darwin at his home in Beacon Street, Lichfield. Hutton rode there with another leading member of the Lunar Circle, the Anglo-Irish writer Richard Lovell Edgeworth. Like Darwin, he was an amateur inventor, claiming to have developed a machine that would calculate the area of a piece of land and caterpillar tracks, which he called 'a cart which carries its own road'. They spent several days being entertained by Darwin, a rotund and jovial host, and then set up an experiment

* It changed its name to the Lunar Society in 1775.

which would have appealed to Hutton's sense of wonder and play-fulness. Darwin later described it in a paper delivered to the Royal Society in London:

> When Dr Hutton of Edinburgh, and Mr Edgeworth of Edgeworthtown in Ireland, were with me about twelve or four-teen years ago, the following experiment, which had been proposed by one of the company, was carefully made. The blast from an airgun was repeatedly thrown on the bulb of a thermometer, and it uniformly sunk it about two degrees. The thermometer was firmly fixed against a wall, and the air-gun, after being charged, was left for an hour in its vicinity, that it might previously lose the heat acquired in the act of charging; the air was then discharged in a continued stream on the bulb of the thermometer, and the event shewed, that the air at the time of its expansion attracted or absorbed heat from the mercury of the thermometer.[6]

Hutton spent six weeks in the West Midlands and from Darwin's home in Staffordshire journeyed north to explore the geology of Derbyshire. He was still a member of the management committee of the Forth & Clyde Canal Company, so it is possible that he was also persuaded by Darwin to visit the Trent & Mersey Canal, then under construction. Like the Scottish canal, it was intended to link the ports of the east and west coasts, enabling vessels to travel between Hull and Liverpool. Darwin had been one of the promot-ers along with Josiah Wedgwood, who wanted to improve the trans-port links for his potteries in Stoke. The engineer was James Brindley, who had also worked for a brief period on plans for the Forth & Clyde.

Hutton and Darwin became friends for the remainder of Hutton's life and kept up a long correspondence of which only one letter survives. In May 1778 Darwin's son Charles,* who was a student at Edinburgh University, died. Darwin probably stayed with Hutton at the time of his son's illness and death and later, at the request of Erasmus, Hutton oversaw the selection and inscription of a

---

* Not the naturalist, who was Darwin's grandson.

gravestone.[7] In the same letter Erasmus gave Hutton medical advice, blaming a decaying tooth for Hutton's headaches and giddiness. He also urged Hutton to take regular meals but warned: 'vinous spirit from small beer to alcohol destroys us all'. Although the letter ends with an exhortation to Hutton and Black to visit him, Hutton never returned to the Midlands.

Hutton left Birmingham at the beginning of July 1774 to set out on his own to see more of Britain and its natural features. From Bridgnorth, Shropshire – presumably his first overnight stop – he wrote to George Clerk Maxwell wishing he had a companion, such as James Keir, whom he had met in Birmingham and had previously been a captain in the army:

> this being the first day that I have begun to be friendless I think it is time to look some out with the eye of faith – I left Birmingham today – have been in Derbyshire & am now in Bridgenorth [sic] with my arse to the east & face to the Irish channel being willing to see thro' Wales or at least to look at it; it will cost me some leather I doubt – I propose steering due west till the sea or something else take me up – I wish I had the Capt for a companion in this voyage. I suppose I shall go thro' this country & not find a soul to ask a question at. He would find a friend in every town & claim kindred to some ancient family in Wales.[8]

The purpose of his trip was to learn about the geography and geology – and he had been given a special mission by Watt to find deposits of a crystalline rock – but as an experienced farmer he could not help commenting on the countryside he passed. He was impressed by the productive enclosed fields and the advanced crop of hay, but was dismissive of the old-fashioned ploughing, which was well behind the modern methods he had introduced on his own farm a decade earlier: 'I have seen, more than once, five great waggon horses yoked endwise like a string of wild geese – doing what? what I could not have believed if I had not seen, drawing two harrows, but for what purpose God knows?'

Hutton's letter style is more a stream of consciousness than a crafted piece of prose and he moves from observations he made

along the route, to details of his personal circumstances and back
again without context or punctuation:

> at Dudley one stage from Birmingham there crops out in a
> ridge coal limestone and whinstone – there are in this country
> spots of Scotland here & there as a lady puts patches on her
> face to make it look like the Divel's [sic] arse – Hay is a full
> crop & very forward they want weather now for making it or it
> will rot in places – wild roses are in blow – I only know the
> season by what I see & feel. I have eat green goose,* but not
> withstanding the weather would seem to favour spontaneous
> generation.

Hutton spent approximately three weeks journeying through the
West Midlands, Wales and then into Wiltshire and Somerset.
Although at least once he took a 'chaise' – a light, fast carriage
pulled by one or two horses – he was mostly riding on horseback
and saddle-sores took a toll both on his body and his trousers. When
he arrived at the Pelican Inn, Bath, he recounted his experiences in
a second letter to Clerk Maxwell, likening his journey to that of the
Children of Israel, fleeing from Egypt to the promised land:

> I wrote you as I was going into the wilderness. I now write
> from the midst of the land of Canaan where have safely arrived
> after as much tumbling & tossing as might have composed an
> *Anead Huttoni* [an Aeneid, an epic poem] – now you cannot
> say but I'm a beast of some bottom and I begin to think there
> is more bottom than head – had my head been beaten as my
> bottom has – and I'm sure it has much better deserved it – lord
> pity the arse that's clagged to a head that will hunt stones – had
> the children of Israel travelled 40 year as I have done they had
> not had a rag to covered [sic] their nakedness – a backside may
> grow – as mine can testify after loss of substance; but how a
> pocket or a breek [trousers] should not wear when mine in
> traveling less than 40 days has been more than 4 times mended

---

* Young goose, traditionally eaten at Whitsun, which had occurred when
Hutton and Watt were en route from Edinburgh to Birmingham.

– my arse it is evident is now a part of much greater consequence than my head*, for a barber never looks at me whilst
here a taylor [tailor] is my constant companion – I carried one
with me in a chaise to Wiltshire and a Taylor [tailor] is the only
acquaintance I have at Bath – no joke, dead ernest [sic].[9]

Hutton goes on to muse about his solitary days – 'It is a great advantage to a naturalist to have no acquaintance in a place provided he
is to stay some time for he must walkabout & enquire to prevent
hanging himself thro the day and then at night he writes a bit and
drinks a soup [sup] of quiet toddy – the world – the world is to him
no more than a turnip.' Hutton has sometimes been portrayed as
teetotal and Playfair says he drank no wine, but from these letters it
is clear he liked a drink before bed, ending his previous letter to
Clerk Maxwell: 'I have just mudled [sic] with brandy & water & so
to bed.' The letters also give us a hint about Hutton's attitude to
women. At home in Edinburgh he appeared as a convinced bachelor, living with his maiden sisters,[10] but he clearly missed the
company of women, writing to Clerk Maxwell:

I begin to be tired speaking to nothing but stones and long for
a fresh bit of mortality to make sauce to them like – we cannot
have everything exactly to ones mind but if there is to be an
excess I think I would chuse it to ly upon the side of quietness
& silence only that we shall have enough of it in the grave
before the resurrection begins mayhap, so let us speak a little
before we ly down – here is the last drop of my sixpenny'th of
toddy to *omnibus freindibus concubinibus ubicumque*; they are
not *hic* they must be *alibi* and *forsan nullibi*; notwithstanding
we may drink them nevertheless.[11]

The mock Latin translates approximately as 'Here's to all concu-
bines wherever they are; they are not here, they must be elsewhere
or perhaps, nowhere.' His first letter had ended with a phrase which
might suggest he had sought more physical comfort: 'a slice of
Cucumber is A that I have got in the vocable of C and that you

---

* This suggests that Hutton was bald by this time. He was 48.

know is no provocative.' The scholars Jean Jones, Hugh Torrens and Eric Robinson, who first published these letters in 1994 believed that the phrase meant that Hutton had had sex with a married woman – a slice of cucumber being the equivalent of the slang expression 'a slice from a cut loaf'.[12] Readers can make up their own minds.

While at Bath, Hutton also had an exchange of letters with James Watt, in which the engineer told Hutton that during his absence he had solved the problems preventing his rotary steam engine from working efficiently. Hutton in turn reported on his quest for Watt to find deposits of a 'crystalline' rock. It is unclear what this substance was or why Watt needed it. He had been a shareholder in a Glasgow porcelain factory and so might have been interested in deposits of petuntz, a type of feldspar used in its manufacture, or he might have wanted one of the minerals used in making flint glass at Keir's works at Stourbridge. Boulton and Watt sold some of Keir's products. Hutton answered in his usual elliptical style:

> I have found the crystalline, but where I least expected – I could not help singing all that day long – what do you think was the song – 'she sought him east, she sought him west, she sought him broad & narrow. She sought him in the cleft of rocks, she found him drown'd in Yarra.'*

He added: 'I wish it had been in your arse it would have saved mine many a knock – nevertheless I have not returned without my errand.'[13]

Hutton by now was known as a fossil collector and he used his trip through Wales and the West of England to increase his collection, writing to Clerk Maxwell: 'I have this day packed a hogshead of bibles all wrote by Gods own finger,' which he was sending back to Edinburgh via Greenock – presumably through Bristol and then by sea. He had also answered one of his own queries from his canal work, whether there was a cheap domestic alternative to imported cement: 'I have got full confirmation in a conjecture that I always

---

* A quote from the traditional Scots song 'Rare Willie Drowned in Yarrow'.

had with regard to the limestone that sets in water. It is just a marly limestone – I used to say that some of our own mers beds would do very well; now I am almost sure of it.' But this part of his geological journey was nearly done. 'I think I know pretty well now what England is made of, only a bit of Cornwall [is left]. I should be glad to see but that I cannot do now for my money will not hold out.'

But Hutton's desire for travelling and adding to his geographic and geological knowledge was not satiated. He had brought funds with him, but they ran out before his exploratory zeal. On his return to Birmingham he borrowed cash from Boulton, which he later refunded – although Boulton had called it 'a favour not to be repaid' – by sending him foodstuffs which could be sold: 'a boll of oat[s] as much of pease & double that of barley meal, with the prudent management of which a good deal of business may be done'.[14]

After a brief visit he set off again for Shropshire and north Wales, climbing the Wrekin, a 1,300-foot (400-metre) hill with Charles Greville, botanist, mineralogist and politician. They reached the top in darkness and 'groped their way home by the light of the stars'. They found it to be made of rhyolites and quartzites.[15] Later Hutton visited a copper mine in Anglesey. He was impressed by the efficiency of the operation, spoke to the manager and was fascinated by the process. In Manchester he wrote again to Watt. 'I have left Wales and shook the dust off my feet upon that occasion – upon the holly head [Holyhead] road is nothing to be met with but either vultures or bad doing. If you travel in a machine they pick the inside of your purse if on horseback you peel the outside of your arse. Glad was I to feel myself again even in England.'[16] He made a brief detour to Rochdale to try to see the immense collection of shells, fossils, stuffed birds, and tribal artefacts amassed by Sir Ashton Lever, but was disappointed: Lever was packing them up for transportation to London.

We have little detail of the rest of his stay in England, other than it involved a lot of partying. He subsequently told Watt:

I made a lucky escape from Warwickshire but in avoiding Scylla I fell into charibdes [Charybdis].[17] I was as tired of eating & drinking as a bachelor is of fasting and a married man of kissing at home . . . In despair I made a violent struggle but

getting out upon the wrong side 1 went to Buxton as the likest
place to Scotland; there I had the mortification to find that had
I not been so engrossed with the beastliness of gluttony and the
manliness of drinking I might have met my friends at Buxton
with Omiah who was there only 8 days before.[18]

One of the friends at Buxton was probably the explorer and
naturalist Joseph Banks, whom Hutton had met in 1772 in
Edinburgh. Omia (spelt by Hutton with an 'h') was a Tahitian
whose arrival in Britain in 1774 made him an instant celebrity.
Banks took him under his wing and was showing him around the
Midlands, using Overton Hall, his family home, as a base.

Hutton's journey home to Edinburgh involved – he told Watt,
rather tongue in cheek – 'the most amazing hardships', and no
sooner was he home than he received an 'incendiary letter from
Madam Young' – possibly the widow of his former apprentice
master Dr George Young, who had died in 1757 – summoning him
to stay with her in Perthshire 'at the foot of the Grampians'. His
letter to Watt on his return home is full of his usual jokes and innu-
endo, but still betrays an affection for the younger man:

Now that I am returned to the empty place where my friends
and yours should be so you cannot expect to hear anything
from a place where nothing is to be found – I write this to
desire something to fill my vacuum, which nature you know
abhors – this is the reason that philosophers whose business it
is to turn nature upside down have invented cylinders full of
steam with condensers at their arse which is vexing whipping
and spurring nature to work out of her ordinary course that
these bougres [buggers?] may sit idle on their arse and look
frae them.

Not so St Samson who was a holy man tho neither a philosopher nor bachelor, god knows, he turned the mill himself and
after lying all night with a whore at Gaza he carried away the
gates of the toon [town] next morning on his back and all this
without any subterfuge or second-hand work – a modern
gentleman is not satisfied with simple action & reaction but
when he goes to bed must have elasticity forsooth to work for

him: what fruit can be expected from such operations? Is not this the true source of the degeneracy of the age? – A smith here has been consulting me about taking out a patent for some improvement of a bed; I'm thinking of adding to it a machine which shall be called the muscular motion, whereby all the several parts shall be performed of erection, intrusion, reciprocation and injection; this will become absolutely necessary in Christian countries that do not allow the eating of children[19] and where people will have pleasure at the easiest rate.[20]

Up until this time Watt had not earned a penny from his inventions and his canal work had ended with the financial crash of 1772. His wife had died in childbirth in 1773, leaving him with two young children to support. His first engine was essentially a water pump, producing a vertical motion, but Boulton was convinced that there was a much bigger market for a machine which could provide rotary power, necessary for cotton and woollen mills and engineering firms. To replace the horses and waterwheel in his own factory he had tried to produce a working prototype steam engine in collaboration with Benjamin Franklin[21] and others were working towards models which could be patented. There was a race to solve the engineering and manufacturing problems. Hutton was concerned that Watt's naïvete and openness would lead to him being exploited, his ideas stolen and him being denied the financial benefits of his inventions.

He was not convinced that Watt's steam engine could be financially successful and spent some time trying to find him a job in Scotland. George Clerk Maxwell had several public appointments and Hutton hoped that he could find a post for Watt, although that would mean giving up his independence in exchange for financial security. His letter continued:

you have not been out of mind tho out of sight this winter – your friends are trying to do something for you – what success will attend their endeavours time only will shew ... come & lick some great mans arse and be damn'd to you; what signifies blowing wind in the arse of a hot barrel, or making soap bubbles in another way than that you did at school.[22]

But in the same letter he appears to accept that Watt is unlikely to return to Scotland and accept some official sinecure. So he urges him to seek protection from parliament for his innovations.

> *Mes Amis*, I wish you all a happy year – may it be fertile to you in lucky events but no new inventions! Invention is too great a work to be well paid for in a state where the general system is to be best paid for the thing that is easiest done. No man should invent but those that live by the publick. They may do it thro gratitude and those who from pride chuse to leave a legacy to the publick; every other man should invent only as much as he can easily consume himself & serve his friends. It is in this view that I admire so much your reciprocating engine . . .
>
> Therefore application should be made to parliament and give the choice either of buying the publick use and description of the invention or of giving an exclusive privilege for a sufficient length of time.[23]

Watt suffered another blow a few months later when William Small died unexpectedly at the age of 40. Like Hutton, he had been a close friend and supporter of Watt, but had more faith that the steam engine would prove successful, telling him that he 'hoped soon to travel in a fiery chariot of your invention'.[24] Hutton sent Watt a letter of condolence, but he had still not given up finding him a job, lobbying Clerk Maxwell and other friends. Failing Watt's return to Scotland, he urged him to go to London and demand that if the government wanted him to remain in England they should find him a job to provide for his family. Watt's replies to Hutton do not survive, but he did not return to Scotland, nor take an official job in the south. Although it took a lot of lobbying, Watt and Boulton took Hutton's advice and obtained an exclusive patent on their steam engine design.

Chapter 11

Look with a kinder eye towards me

IN 1778 HUTTON received a letter from his son. The younger James Hutton was 29 or 30 years old and working as a junior clerk at the General Post Office in Lombard Street in the City of London. We believe he grew up in London, although where he received his education or what his first employment was, we do not know. He entered the postal service in 1770 as a 'junior facer of letters',[1] working on the 'West Road' – presumably serving the west of England and Wales. The British postal service, one of the most sophisticated in the world at the time, was a lucrative source of income for the government of the day and for the aristocrats who were appointed as one of the two postmasters general. The distribution of the mail from London was divided into 'roads' each run by a clerk, who had responsibility for the revenues and costs incurred in the management of his part of the service, including the payment from the profits of pensions to previous employees at levels set by the board. Hutton's chance of an appointment came after a reorganisation of the West Road following the death of the clerk.

At the same time George Oxlade, an assistant to the West Road, was 'allowed to resign' and given an extra boost to his pension – at the expense of more junior staff – in recognition that his salary had 'not been adequate' for his long service. On Oxlade's departure, the 'next in rotation' took over as assistant, leaving a gap at the bottom. We don't know what James Hutton junior was paid, but salaries for staff at the lower levels were clearly not high and some had been used to supplementing their income elsewhere. The Post Office board was keen to stop this practice. The entry above that detailing James Hutton's appointment reads: 'ordered that no clerk, letter carrier or other person employed in this office do enter into any trade or business without leave first obtained of this board'.

In 1775 James had married Alice Smeeton in St John's Church, Wapping, London.[2] Alice had been born and brought up not far away in Stepney and was 18 at the time of her wedding. Because she was not 'of age' but had the consent of her parents, the couple had to obtain a licence to marry, rather than calling the banns for three weeks, which was the more normal procedure. The witnesses were Jane Smeeton, Alice's sister, and Elizabeth Thomas, about whom we know nothing. It appears to have been a love match and James was quickly welcomed into Alice's family. When her father William made his will two years later he described James as 'my beloved son'. Within two years the newly married couple had two children, William, named for Alice's father, and Margaret. They gave each the double surname Smeeton Hutton.

Hutton's father-in-law died in December 1777, leaving six houses in Wapping and Stepney to James, Alice and his other daughter, Jane. He did not own them all, some were leased, but all the properties were let to tenants. This was less of a windfall than it appeared – the rents were not only for his daughters and son-in-law, they had also to provide an income for his widow, Barbara.[3] The buildings were in a poor state of repair, so the rents which could be charged were low. By the time the lease payments had been subtracted there was little left and James' earnings from the Post Office were not sufficient for him to be able to redevelop the properties. He had another problem: the rented home he and Alice occupied was cramped enough for a couple with two children, but in the summer of 1778 Alice became pregnant again.

James began thinking of a plan. If he could borrow money, he could build a bigger house for himself and his family and repair and improve the other properties, enabling him to charge higher rents. He sketched out the figures and estimated that he needed to borrow £500 – a huge sum compared to his own salary. Britain was at war with the American colonies and to finance it the government was issuing large amounts of debt causing interest rates to rise sharply. To compete with the return investors could get on government stock, James would have to offer annual interest of at least 5%.[4] He calculated that with the extra income and the saving on the rent he and Alice were already paying on their home, he could afford it, but where to find men willing to lend to a lowly clerk? He did not move in the social circles where he could meet people with that amount of spare cash. He wrote to his father.

That letter and James Hutton senior's reply have not survived, but from a second letter James junior sent in October that year we can piece together the story.[5] James wrote to his father asking for recommendations for men he might approach for the loan. What he got in return was a 'bond' endorsed by his father and John Davie, Hutton's partner in the chemical business. This would provide the security that the younger man needed to borrow for his scheme. He was overwhelmed:

> When I wrote to you my hopes of improving my income by rebuilding some old houses and requesting your assistance I meant your recommendation to your monied acquaintance as I had none myself. I neither desired nor expected to have it directly from you. I hope to have you always look with a kinder eye towards me than the strict laws of trade would admit of. I wish'd and indeed founded my chief hopes of having it from your friends in London, both as they could be informed of more particulars and see more reasons to be satisfyed [sic] with their security on the spot than I well could commit to in writing.

We don't know if James Hutton junior did go through with his property development. He lays out the calculations for his father's approval, showing that he would be capable of paying the interest. But he appears to have doubts, writing: 'You will pardon me if I give you the reasons I had for hoping that I should increase my income by that scheme, not that I persist in it now, however anxious I am of devising something that might answer my design.' There is a hint of an unwelcome development which might have upset his calculations at the end of the letter: 'Our salarys [sic] are taxed 4 in the pound Land Tax this Michaelmas for the first time and that for half a year past, which makes the deduction five pounds out of this quarterly payment: a deficiency as afflicting as it was unforeseen and unprovided for.'

The letter is important because it shows that Hutton was in touch with his son and anxious to help him when he could. It also shows that John Davie, who allowed his name to be added to the bond, was aware of the existence of the son. Frustratingly we do not know how the relationship between father and son developed, or whether he ever met his grandchildren – there is no more documentary evidence until after James Hutton senior's death. His new grand-child was born on 28 April the following year. He was named James Edington Smeeton Hutton. The reason for the choice of the boy's first name and his double surname are obvious; that for Edington is unclear. Was he merely named after a family friend, or is it a clue to the identity of his grandmother (see Ch. 1)? We do not know.

Another incident from this period of the older Hutton's life rein-forces the idea that far from abandoning his son he would have loved to have been a proper father. Two of his cousins from his mother's side of the family, Andrew and John Balfour, had emigrated to North America. Andrew Balfour left Scotland in 1772, sailing from Greenock but leaving behind his wife and infant daughter Isabel (known as Tibbie). His wife died a year later, and Balfour married Elizabeth Dayton who was also Scottish, but living in Newport, Rhode Island. Failing to earn a living in the north, Andrew joined his younger brother, John, in Charleston, South Carolina, where they set up a salt manufacturing business. In 1778 Andrew moved to North Carolina, where he bought a plantation and sent for his daughter who made the journey from Scotland with her aunt Margaret, Andrew and John's sister. John stayed in South Carolina where he also bought a plantation. At the outbreak of the rebellion against British rule the two brothers took opposing sides. John remained a loyalist, but South Carolina saw heavy fighting and he was killed in November 1781 when his house was burnt down. Andrew became an officer in George Washington's Continental Army. He was a close friend of Washington and fought in Georgia and the Carolinas. Four months after his brother's death, while he was home on leave, Andrew Balfour was surprised by a loyalist raiding party and killed in front of his sister and daughter.[6]

When he heard the news James Hutton wrote to both widows offering support if they wanted to return to Scotland. Andrew's widow, Elizabeth, opted to stay in America,[7] but John's widow, Mary

Ann, decided to return with her two daughters and a son, also named Andrew, who was less than a year old. Hutton supported the family financially and they may have lived with him and his older sister Isabella for a while. He took a particular interest in the education of young Andrew Balfour, who 'was often said to have accompanied Dr Hutton on his rambles round Arthur's Seat and Kings Park'.[8] Hutton may also have influenced Balfour's decision to study medicine at the University of Edinburgh.[9] Andrew later married and had a son, born in 1808. He named him John Hutton Balfour, in memory of his father and affection for his cousin, who had acted as surrogate father to him. After the death of Isabella Hutton in 1818, Andrew, then 37, moved his family into Hutton's house in St John's Hill, Edinburgh.

The gratitude and affection shown by Andrew Balfour to Hutton suggest that, had he been able to do so, Hutton would have wanted to play a part in the upbringing of his son. This makes his estrangement even more mysterious.

★ ★ ★

Adam Smith moved to live in Edinburgh in 1778 to take up the post as a Commissioner of Customs. He was already an intellectual celebrity through the popularity of his two books, the *Theory of Moral Sentiments* (1759) and *An Inquiry into the Nature and Causes of the Wealth of Nations* (1776), which was to cement his reputation as an economist as well as a philosopher. He bought a large town house just off the Canongate, which had been built in the late 17th century for the Earls of Panmure, and lived there with his mother, his spinster cousin, Miss Jean Douglas, and her nephew. It was less than ten minutes' walk from Hutton's home in St John's Hill.

Hutton and Smith already knew each other. They were perhaps introduced during one of Smith's visits to Edinburgh by Joseph Black who had been a colleague of Smith's at the University of Glasgow. He had also been the physician to Smith's close friend David Hume, who had died two years previously. Hutton had written to Smith when the economist was living with his mother in Kirkcaldy working on his book. It was a brief and enigmatic letter:

I send you this flax in the ear to inform you that November is begun and that there is now little danger of frost until after the

New Year; so if you have anything to do with what is without you may conduct yourself accordingly; if it is otherwise and you are made for sleep and vision let me know when I should awaken you again.[10]

Smith had many friends in Edinburgh. He had delivered a series of lectures at the university between 1748 and 1751 at the request of the judge Lord Kames, who was a founding member of the Philosophical Society and patron to some of the city's leading thinkers. Smith had kept in touch with many of the people he met, including William Robertson, who had become principal of the university, and Adam Ferguson, who was the professor of moral philosophy. Yet for the last dozen years of his life, his closest friends were James Hutton and Joseph Black.

Smith was a much more worldly character than either Hutton or Black. Where they were interested in the natural world, his work concerned the material relationships between people. He was a friend of aristocrats like the Duke of Buccleuch, to whom he had acted as tutor and who had provided him with a pension. He was consulted by government ministers on financial matters and he took his official position as one of the collectors of customs duties very seriously. He was not dour, but did not have the humorous playfulness of Hutton and he could be absent-minded. Occasionally he would be so deep in thought while walking that he ended up in a place he did not intend to go; or when signing a document copied the name of the person who had signed before him, rather than writing his own. An early biographer of Smith believed the different characters of the three friends knit them closer together:

Black was a man of fine presence and courtly bearing, grave, calm, polished, well dressed, speaking, what was then rare, correct English without a trace of Scotch accent, and always with sense and insight even in fields beyond his own. Smith used to say that he never knew a man with less nonsense in him than Dr Black, and that he was often indebted to his better discrimination in the judgment of character, a point in which Smith, not only by the general testimony of his acquaintance, but by his own confession, was by no means strong, inasmuch as he was, as he acknowledges, too apt to form his opinion from a single feature.

Hutton was in many respects the reverse of Black. He was a dweller out of doors, a man of strong vitality and high spirits, careless of dress and appearance, setting little store by the world's prejudices or fashions, and speaking the broadest Scotch, but overflowing with views and speculations and fun, and with a certain originality of expression, often very piquant.[11]

Hutton and Smith sometimes walked and talked together. One such occasion produced an anecdote which Hutton told to the novelist Henry Mackenzie:

> On their walk one day in Leith, Smith suddenly stopped in the course of one of his reveries and looked earnestly on a corn-field, which had been wheat but had been reaped some months before. Hutton, in joke, said: 'That's a famous crop, is it not?' 'Why yes,' replied Smith, seeing the crop in his imagination, 'and it is not difficult to account for it. The vicinity of the town furnishes abundance of manure and the application of labour is now better understood than formerly.' Dr Hutton pursued the Joke no further & they pursued their walk without any further mention of the crop.[12]

The Scottish sabbath was not yet observed as strictly as it later became, and Smith began a series of Sunday evening dinners at his home. Hutton and Black were regular attenders. Conversation and good-natured argument were the main attraction of these events, rather than the food and drink. Smith's biographer adds: 'Probably no three men could be found who cared less for the pleasures of the table. Hutton was an abstainer;* Black a vegetarian, his usual fare being some bread, a few prunes, and a measured quantity of milk diluted with water; and as for Smith, his only weakness seems to have been for lump sugar.'[13] Smith's fondness for sugar and his absent-mindedness gave rise to another anecdote: 'The author of the *Wealth of Nations* never thought of marrying. His household

---

* This was not strictly true, although Hutton did not drink wine.

affairs were managed to his perfect contentment by a female cousin, a Miss Jeanie Douglas, who almost necessarily acquired a great control over him. It is said that the amiable philosopher, being fond of a bit of sugar, and chid by her for taking it, would sometimes, in sauntering backwards and forwards along the parlour, watch till Miss Jeanie's back was turned, in order to supply himself with his favourite morsel.'[14]

The three friends also started the Oyster Club, which met every Friday at 2 p.m. at a tavern in the Grassmarket, below the castle rock. Clubs were a feature of 18th-century intellectual life and Edinburgh had many. Smith, even during the period when he was a professor at Glasgow University and a member of at least two Glasgow clubs, made the long and uncomfortable coach journey to Edinburgh to attend meetings of the Select Society. Its highly influential membership included David Hume, the painter Allan Ramsay, and Alexander Wedderburn, who later became lord chancellor.[15] Black had been a member of the Poker Club, which had started life as the Militia Club, intended to campaign for the establishment of a citizen army in Scotland to guard against attempted coups like the Jacobite rising of 1745–6 and generally to safeguard liberty. England's right to recruit a militia was established in an Act of Parliament of 1757, but a similar bill for Scotland failed. The name provoked some opposition, so at the suggestion of Adam Ferguson it was changed to 'Poker' – a reference to poking a fire, rather than the card game. Black's membership of a club concerned with the protection of liberty, suggests that his reactionary remarks to the young Silas Neville (see Ch. 9) were tongue in cheek. Hutton was never a member, perhaps because of his lack of interest in politics.

At its height the club attracted many of the leading names of the Edinburgh Enlightenment. The theologian Dr Alexander Carlyle, who kept the minutes, described the management of the Poker Club as 'frugal and moderate'. Members met at Thomas Nicolson's Tavern, were served a dinner costing a shilling a head at 2 p.m. and called for the bill at 6 p.m. Sherry and claret were served, but 'no approach to inebriety was ever witnessed'. Nevertheless, an altercation between some of the members and the landlord eventually led to the club having to meet in the much more expensive Fortune's Tavern and it dwindled away.[16]

The Oyster Club appeared to have a less formal structure than the Select Society, the Poker Club, or the Philosophical Society. If written rules existed, or minutes were taken, they do not survive and there is no record of officers being elected, as there were for the other associations. It appears to have been an excuse for friends to meet and talk, with no overtly political or academic agenda. Besides Hutton, Black and Smith, the gathering attracted the lawyer and novelist Henry Mackenzie, the mathematicians Dugald Stewart and John Playfair, Sir James Hall, the son of Hutton's university friend Sir John Hall, architect Robert Adam and John Clerk of Eldin, who had studied anatomy with Hutton. Franz Swediaur, an Austrian doctor, who spent some time in Edinburgh in 1784 researching with William Cullen, was made a member of the club during his stay: 'We have a club here which consists of nothing but philosophers ... Thus I spend once a week in a most enlightened and agreeable, cheerful and social company.'[17]

No new scientific discoveries came out of the club's meetings, but its valuable legacy to history was the friendship between Hutton and Playfair, which led the young mathematician to write the first biography of Hutton[18] – the basis for all subsequent accounts of his life, including this one. According to Playfair, the chief delight of the club was to listen to the conversation of its three founders:

As all the three possessed great talents, enlarged views, and exten- sive information, without any of the stateliness and formality which men of letters think it sometimes necessary to affect, as they were all three easily amused, and as the sincerity of their friendship had never been darkened by the least shade of envy, it would be hard to find an example where everything favourable to good society was more perfectly united, and everything adverse more entirely excluded. The conversation was always free, often scientific, but never didactic or disputatious; and as this club was much the resort of the strangers who visited Edinburgh, from any object connected with art or with science, it derived from thence an extraordinary degree of variety and interest.[19]

The tavern in the Grassmarket might have been the second venue for the Oyster Club. The caricaturist John Kay tells the story of

Hutton and Black choosing a hostelry on South Bridge for the initial meeting place.

> Without further inquiry the meetings were fixed by them to be held in this house and the club assembled there during the greater part of the winter, till one evening Dr Hutton, being rather late, was surprised when going in to see a whole bevy of well-dressed, but somewhat brazen-faced young ladies brush past him and take refuge in an adjoining apartment. He then, for the first time, began to think that all was not right and communicated his suspicions to the rest of the company. Next morning the notable discovery was made that our amiable philosophers had introduced their friends to one of the most noted houses of bad fame in the city.[20]

Hutton was also introduced to Yekaterina Dashkova, a Russian princess who was living in the Palace of Holyroodhouse while her son Pavel attended the university from 1776 to 1779. She was a woman of learning and taste, having studied mathematics at the University of Moscow and been received at the courts of many European countries before reaching Scotland. In Paris she had befriended the philosophers Diderot and Voltaire and she had corresponded with William Robertson, principal of the University of Edinburgh. She was received warmly by leading members of the Edinburgh Enlightenment, including Adam Smith and Adam Ferguson, but despite her learning and wide interests she would not have been able attend meetings of the Oyster Club or any other of the intellectual clubs and societies, which did not admit women. Instead she entertained them regularly at the palace. There is no record of Hutton attending any of these soirées, but at Joseph Black's suggestion he took on the task of organising and cataloguing a collection of Derbyshire fossils she had collected while on a visit to take the waters at Matlock and Buxton.[21]

Chapter 12

A Junto of Jacobites and Tories

THE LAUNCH IN 1783 of the Royal Society of Edinburgh (RSE) gave Hutton a position at the centre of the intellectual life of the city and a platform from which to launch his ideas and expose them to critical scrutiny. He became a founding fellow, joined the council and was one of the first to read a paper at its meetings. Had the society not been formed we might know a lot less about Hutton's work and our view of the development of the Earth might have been different. He was an active member for the rest of his life, but his influence persisted long after his death. Over the next half-century the RSE became a bastion of support for Hutton's theories against sustained attacks from men with much higher academic standing than he possessed. Yet the establishment of a national academy for Scotland to rival the Royal Society in London or the Académie Française in Paris was not inevitable, rather it was the product of bitter personal and political rivalries and a clash of huge egos, to which Hutton was either an uninvolved observer, or more probably, unaware.

By the end of the 1770s the Philosophical Society was losing momentum. It had published nothing since the last volume of papers in its 'Transactions' series in 1771 and Lord Kames, the driving force in the previous decade, was ageing and ailing. Younger members believed that the society was paying too little attention to science and had become overly bureaucratic.

In 1778 a group of leading professors and doctors who, with one exception, were in their thirties, formed the Newtonian Club, as a faction within the society. It adopted just six very terse rules which tell us a lot about the parent body. Rule 1 stated: 'That as a multiplicity of laws has a direct tendency to produce confusion instead

of order, it is resolved to limit their number as much as possible.'
Rule 6 said: 'That as this club consists entirely of philosophers, it
would therefore be ridiculous to make any laws for its internal
police.'[1] We don't know in detail what the 'internal police' might
have been doing in the Philosophical Society, but as far as its more
radical members were concerned, it was not encouraging open
debate and furthering the spread of knowledge. There may also
have been a political dimension to the move: some of the Newtonian
Club members were known to have Whig sympathies and to favour
electoral reform, whereas Kames was a staunch Tory. Two of the
founding members of the club had been rejected for professorships
at the university, which was solidly under Tory control.[2]

A second, this time external challenge to the Philosophical
Society came in 1780. David Erskine, the 38-year-old Earl of
Buchan, called a meeting in his Edinburgh home to propose a new
organisation, the Society of Antiquaries of Scotland. Buchan was a
precocious and opinionated man. He had been elected a fellow of
two prestigious London bodies, the Royal Society and the Society
of Antiquaries, but his membership lapsed shortly afterwards for
non-payment of fees. He was a freemason and served as the Grand
Master Mason of Scotland and he was a campaigner for political
reform. He came from one of the leading Whig families of Scotland
and had made speeches condemning the handpicking by the Tory
lord advocate Henry Dundas of the peers who were to represent
Scotland in the House of Lords. He was described by Sir Walter
Scott as 'a person whose immense vanity, bordering on insanity,
obscured, or rather eclipsed, very considerable talents'.[3]

Buchan told the 14 prospective members who accepted his invi-
tation that his motivation for forming the new society was because
there was 'no regular Society for promoting Antiquarian researches'.[4]
This wilfully ignored the historical papers which had been read to
meetings of the Philosophical Society, but the study of Scotland's
past was not the sole object of his interest. In a letter to William
Smellie, whose passion was natural history, Buchan added that 'it is
meant to widen the field of enquiry to pursuits connected with it,
whether natural, moral, or political'.

Smellie was by trade a printer, but he had studied botany under
John Hope, the professor at Edinburgh University, and when Hope

was away had taken his place to deliver lectures. In his career as a publisher he produced works by David Hume, Adam Smith, William Robertson, Adam Ferguson, Joseph Black, James Hutton, Lords Hailes, Kames, and Monboddo, and Robert Burns, among others. He was the first editor of the *Encyclopaedia Britannica* and wrote several of its entries himself. He had written and published a pamphlet calling for the reform of the judicial system and another attacking the power of Henry Dundas. He spent most evenings in oyster bars and ale houses, such as Douglas's Tavern where he founded the Crochallan Fencibles, a drinking club immortalised in the verse of his close friend and political ally Robert Burns, whose obscene *Merry Muses of Caledonia* was written for the club.[5]

Smellie was a member of the Philosophical Society and the Newtonian Club and when the position of professor of natural history at the university had become vacant in 1775 he applied. His opponent was Dr John Walker, a Church of Scotland minister serving the parish of Moffat, 50 miles from Edinburgh. He was a close ally of William Robertson, principal of the university, but also a noted botanist, who had made an extensive survey of the Highlands. Smellie and Walker were on opposite sides of one of the big scientific debates of the age: Walker was a supporter of Carl Linnaeus, the Swedish academic who had invented a method of classifying and naming plants. Smellie rejected Linnaeus and published his own theory, which Walker ridiculed in an article in the *Edinburgh Weekly Magazine*. In the contest for the professorship, Walker won, but under the auspices of the Society of Antiquaries of Scotland, Smellie was to be given a chance to get his own back.

Buchan invited Smellie to give a series of public lectures on the 'Philosophy of Natural History'. This was a direct challenge to the authority and earnings of Walker at the university, whose stipend was meagre and who depended for most of his income on the fees paid to attend his lectures. Buchan also announced that the hall he had bought for the society in the Cowgate would become home to a museum, of which Smellie would be the keeper. This was another swipe at the university which had no collections of any value. The previous professor of natural history had been given botanical and mineral samples – some donated by the Earl of Buchan himself – but after his death the trustees of his estate successfully argued that they were personal gifts and removed them from the university for sale.[6]

The last straw for Walker and the university came in 1782 when Buchan petitioned King George III for a Royal Charter for his society. The only organisations in Edinburgh with this Royal seal of approval were the medical societies. Unlike London or Paris, there was no national academy; the university was the creation of the council, which was under the political direction of Henry Dundas, and the Philosophical Society was a self-perpetuating body of uncertain legal standing. A campaign to block Buchan's ambitions was quickly mounted and a paper written by Walker became the basis for a counter-petition from the university calling for a new body to be set up which would subsume both the Philosophical Society and Buchan's Society of Antiquaries of Scotland. The Faculty of Advocates, whose library was to go on to be the foundation for the National Library of Scotland, also joined the conspiracy. The university stated that the king, acting through his Scottish minister, Henry Dundas, would appoint the president and members of the new body – it would be under firm Tory control.

Dundas, however, did not want to be drawn into an acrimonious public scrap and asked Buchan and university principal William Robertson to agree a compromise. The meeting which took place in Buchan's home did not go well. It started with Robertson, whom Buchan insisted on calling 'the historiographer', a reference to his official title as Historiographer for Scotland, reading 'a long memorandum prepared by himself', which outlined his plan to take over the Society of Antiquaries, backed with an implied threat that if Buchan did not agree voluntarily Dundas would compel him. Buchan was no stranger to threats himself, having suggested that the university might be subjected to a Royal visitation – an official audit which could expose corruptions such as professors accepting their stipends and then doing no teaching. Buchan's counter-arguments carried no weight with Robertson, so he resorted to insult. According to Steven Shapin, who uncovered the machinations which led to the formation of the Royal Society of Edinburgh:

The Earl went on to impugn the family origins of both Henry Dundas ("the younger son of a branch of a Private Family . . .

without information, and bolstered by impudence and scurrility") and William Robertson ("an obscure Priest, the brother of an obscure Goldsmith in Edinburgh"), to call their colleagues "a Junto of Jacobites and Tories who insult the best men in Scotland, and determine the existence of Literary Societies, Militias, Armaments and Constitutional Rights."[7]

Fierce lobbying by both sides carried on for a few more months until in March 1783 Dundas decided that he had had enough and recommended to the king that two Royal Charters be granted, one to the Society of Antiquaries and another to the proposed new organisation, which was to be called the Royal Society of Edinburgh. Buchan had his Royal Charter and had seen off the threat of being subsumed, but it was a Pyrrhic victory. The Antiquaries carried on for a few years, holding meetings and publishing papers, but membership declined and with it subscription income to the point where it was unable to meet the interest on the mortgage on its premises, which had to be sold. Buchan himself resigned after ten years.

The Royal Society of Edinburgh held its first meeting on 23 June 1783, electing the lord justice clerk, Thomas Miller, the second most senior judge in Scotland, as its president. He had been one of the signatories to the petition to the king, along with other prominent Tories Dundas and the Duke of Buccleuch. Robertson, William Cullen and several other university professors were also named, as was Adam Smith and George Clerk Maxwell, now Sir George, following the death of his older brother. Neither Hutton nor Black were mentioned.

All members of the Philosophical Society were assumed as members. Hutton and Joseph Black were listed as members on August 4, and at the same meeting honorary membership was extended to a number of eminent foreigners – including the American scientist and politician Benjamin Franklin and Louis Leclerc, Comte de Buffon, who had been head of Le Jardin du Roi when Hutton had studied in Paris.[8] Franklin had visited Edinburgh and was an honorary member of the Philosophical Society, but Buffon is not known to have been to Scotland and must have been elected because of his reputation as a scientist and author.

Invitations to join the new society were sent to the lords of council and session (the judges), the barons of exchequer and 'a select number of other gentlemen'. John Robison, professor of natural philosophy (science), at the university was elected general secretary and Alexander Keith, Writer to the Signet (solicitor), as treasurer. It then adjourned for six weeks to draw up a constitution, which divided the society into two classes – the Physical, comprising the sciences, mathematics and 'whatever relates to the improvement of arts and manufactures'; and the Literary, comprising literature, philology, history, antiquities and 'speculative philosophy'. The university and the Faculty of Advocates got their reward: objects of natural history presented to the society were to be deposited in the university museum and books in the Advocates Library.[9]

The Physical Class met for the first time in November with William Cullen, professor of medicine and a close ally of both Robertson and Walker, in the chair. The first paper read was by Walker, on the 'motion of the sap in trees'. At the first meeting of the Literary Class, two weeks later, Dr Robertson presided and the Rev. Thomas Robertson, minister of Dalmeny, read the first part of his dissertation on his theory of inflection in languages.

The foundation of the Royal Society of Edinburgh (RSE) marked a watershed moment for Hutton. Previously his reputation as a scientist rested on the good opinion of his friends and a slim pamphlet on the difference between coal and culm. We know of only one paper to the Philosophical Society – an observation of various tracks of dead grass on Arthur's Seat, the volcanic rock in the centre of Edinburgh. He described them in detail, but was at a loss to explain them.[10] The paper did not appear to have provoked any comment nationally or internationally. He was now in his mid-fifties and living a comfortable life of study and pleasure. Playfair described his typical day:

> his manner of life . . . gave him such a command of his time, as is enjoyed by very few. Though he used to rise late, he began immediately to study, and generally continued busy till dinner. He dined early, almost always at home, and passed very little time at table; for he ate sparingly, and drank no wine. After dinner he resumed his studies, or, if the weather was fine,

walked for two or three hours, when he could not be said to give up study, though he might, perhaps, change the object of it. The evening he always spent in the society of his friends. No professional, and rarely any domestic arrangements interrupted this uniform course of life, so that his time was wholly divided between the pursuits of science and the conversation of his friends, unless when he travelled from home on some excursion, from which he never failed to return furnished with new materials for geological investigation.[11]

This may seem a casual lifestyle, but in the remaining years of his life he was to reveal himself as a serious scientist of considerable breadth and learning, publishing in quick succession papers covering meteorology, philosophy, language, heat, light and fire and his seminal theory of the formation of the Earth.

He also moved from being an Edinburgh eccentric to being a celebrity. He sat for the fashionable portrait painter Henry Raeburn,[12] was caricatured twice by the satirist John Kay, once with a geological hammer and a second time with his friend Joseph Black, and a relief bust of him in a philosopher's toga was produced by the miniaturist James Tassie. His work began to be noticed in London and further afield.

In 1787 he was made an honorary member of the Royal Academy of Agriculture, Paris.[13] This was remarkable because until the first volume of the *Transactions of the Royal Society of Edinburgh* appeared the following year, none of his papers had been published and he does not appear to have either lectured or written on farming at all. It was an honour which he proudly put at the top of his publications next to his medical degree and his fellowship of the Royal Society of Edinburgh – until outbreak of the Revolutionary Wars with France in 1792, when he dropped it. It is possible that a 'separate' – a printed transcription of Hutton's presentation – had circulated prior to publication in the *Transactions* and that this found its way into the hands of Nicolas Desmarest. The two men had both been students in Paris in the late 1740s. After winning his essay prize in 1752, the Frenchman had been employed by the government of France to study best practice in industry and agriculture around the country and abroad and to disseminate the information as widely as possible. In 1788 he was appointed Inspector-General and Director of the Manufactures of France, a position he held until the revolution. Parallel to his official duties he developed an interest in geology and, like Hutton, travelled his native land in search of information. According to his biographer:

> Resuming the rustic habits of his boyhood, he made his journeys on foot, with a little cheese as all his sustenance. No path seemed impracticable to him, no rock inaccessible. He never sought the country mansions, he did not even halt at the inns. To pass the night on the hard ground in some herdsman's hut, was to him only an amusement. He would talk with quarrymen and miners, with blacksmiths and masons, more readily than with men of science. It was thus that he gained that detailed personal acquaintance with the surface of France with which he enriched his writings.[14]

One particular problem interested Desmarest: the nature of basalt – a question that also interested Hutton. In his presentations to the Academy of Sciences Paris in 1769 and 1771, Desmarest proposed that it was volcanic, thus challenging the orthodox theory that it had been laid down by sedimentation at the bottom of a primal ocean. By 1787 Desmarest was president of the Paris Royal Academy of Agriculture and thus in a position to influence who was awarded its honorary fellowships. Had he and Hutton kept in touch since their student days and exchanged geological observations? No surviving documents from either man proves that they had, but it would explain Hutton's honour and the similarity of many of their ideas and interests.

Hutton's first paper, his 'Theory of Rain', a 16,000-word essay read over two hour-long meetings of the RSE in February and March 1784, was the first scientific attempt to describe the physical conditions which cause rain.[15] It built on his interest in the weather, which dated from his days as a farmer when he had tabulated the effects of rainfall and temperature on rates of crop growth. On his tours around Britain he had recorded thermometer readings in springs, reasoning that these would give him a rough estimate of average temperatures. By plotting these against their geographic position he estimated that average temperatures fell by one degree Fahrenheit* for each degree of latitude travelled north. He had also calculated that the temperature fell by the same amount for each 230 feet (70 metres) of altitude.[16]

The theory also drew on his interest in heat, which he had often discussed with Joseph Black, who had invented the concept of latent heat – the heat stored in a body (in this case a mass of air) until released under certain conditions. Although Hutton did not use the term 'latent heat', he argued that the mixing of two moist air masses of different temperatures caused supersaturation and therefore condensation and rain. He illustrated his ideas with a mathematical model, which he represented by a graph of humidity against temperature. In evidence he cited the phenomenon recorded by Pierre Louis Moreau de Maupertuis (1698–1759), the French mathematician, geographer and astronomer who, on an expedition to Lapland to measure the shape of the Earth, had noted that when the door to a warm room was opened and cold air from outside rushed in 'swirls of snow' were formed from the water vapour in the room. Professor John Robison had noticed a similar effect in St Petersburg when a window was opened in a crowded hot room. A 'visible circumgyration of a snowy substance' had resulted. Hutton's paper went on to talk about rain in other parts of the world, including Egypt, the Caspian, Asia, North America and Peru, quoting observations by others as well as his own ideas.

The dissertation attracted attention outside Scotland. Richard Kirwan, an Irish geologist and chemist, mentioned Hutton's theory at a meeting of the philosophical society which usually met in the Chapter Coffee House, London (although on this occasion, 8 December 1786, it was held at the Baptist's Head Coffee House). The society brought together many scientists, philosophers and mathematicians. Boulton and Watt were both members, along with their fellow Lunar Society colleagues Josiah Wedgwood and James Keir. It is possible James Hutton was also a member, although Charles Hutton, professor of mathematics at the Royal Military Academy was also a member, so there could have been some confusion. Attendance at meetings sometimes includes a 'Dr Hutton' and

---

* 1°F = 0.55°C

at other times merely says 'Hutton'. The 'Theory of Rain' had not at that stage been published, so Kirwan must have either seen a manuscript copy of Hutton's lecture, or been told about it. He was not impressed. The minutes state:

> Inquiry being made for news, Mr Kirwan mentioned a disser-
> tation on Rain by Dr Hutton of Edinburgh, who says he mixed
> two Quantities of air differently heated, marking the degree of
> heat to which the mixture was reduced & observing that a
> Quantity of water was deposited. Upon this his Theory hinges,
> but appears not satisfactory.[17]

Hutton had by this time switched his extraordinary range of interests to another subject entirely. His paper delivered on 19 June 1786 outlined the problems of representing spoken language in writing. He identified and examined four different methods of analysing speech for the purposes of classifying languages. But more serious criticism of his propositions on the formation of rain forced him back to meteorology. The Swiss mineralist and meteorologist Jean André de Luc, a fellow of the Royal Society of London and a serious scientist who had conducted experiments on latent heat with James Watt, published *Idées sur la météorologie* in 1787. He took issue with Hutton's central point that when two masses of air of different temperatures are mixed together the humidity of the combined mass is greater than the mean of the two masses had when separate. Hutton answered his criticisms in a lengthy new paper delivered to the Royal Society of Edinburgh. 'This I maintain to be a physical truth,' he declared, 'Mr de Luc refuses to admit it as a rule in nature.'[18] Hutton not only showed his mastery of physics in his refutation, but his language skill as well. De Luc wrote in French and Hutton quotes his arguments without translations, expecting his audience to understand. His answers, however, are in English.

Playfair, writing after Hutton's death, described de Luc's attack as 'vigorous and determined', Hutton 'defended it with some warmth, and the controversy was carried on with more sharpness, on both sides, than a theory in meteorology might have been expected to call forth. For this Dr Hutton had least apology, if greatest indulgence.'

More temper was shown by 'the combatant who has the worst argument'.[19] Hutton's ideas fitted the evidence as it was then known and stood for nearly 100 years until better measurement techniques showed that although the conditions Hutton had described could lead to the formation of clouds, the amount of water which could be condensed in this way was too small to cause rain.[20]

Chapter 13

The operation of subterraneous fire . . .

ACCORDING TO PLAYFAIR, Hutton began formulating his theory of the Earth at least thirty years before he first presented it in public. That would suggest that he began to think about how the landscape around him had been made while he was studying agriculture and managing his own farm. His stay in East Anglia would have given him a graphic illustration of destruction and renewal. The Norfolk coast around Yarmouth, where Hutton stayed for a year, is one of the most dynamic in Britain, constantly changing shape as soil, sand and gravel are carried to the sea by the River Yare. The need to constantly replenish the land's fertility was well understood by local farmers and became a basic part of Hutton's practical farm management.

On his journeys around England, Wales and Scotland he was alert for signs of the changes that had occurred in the landscape and what might have caused them. He descended into mines, he climbed hills, he sought out knowledgeable local scientists and geographers. He had a prodigious memory and Playfair tells us that he kept voluminous papers, although very few of these survive. He also began to build up a mineral library, collecting samples, exchanging some with other collectors both at home and abroad and occasionally buying rocks and fossil-bearing stones that he thought interesting.[1]

But the image of Hutton as the simple farmer developing his sophisticated theory of the Earth by managing the soil undervalues Hutton as a serious scientist, keen not only to make observations and conduct experiments of his own, but to learn from the experience of others and test their conclusions. Like Newton, he could see far because he stood on the shoulders of giants. We know from

Playfair that Hutton discussed his evolving theory with his close friends Joseph Black and John Clerk. (Black incorporated some of Hutton's ideas in his lectures and Hutton drew on Black's discoveries in chemistry and the physics of heat.) But we also know that he was interested in the explanations proposed by others, if only to disagree or disprove them.

The disappearance of his library after his death means that we do not know exactly whose work he had read, but we do know that he was a voracious consumer of books and papers on natural history and the accounts of travellers. He was a close questioner of men like Banks and Lind, who had visited countries he had not had the opportunity to study.[2] His facility with French and his time in Paris had probably also opened his mind to the important work being published by Continental scholars. By the time he started farming he had already received a theoretical grounding: from Rouelle's lectures in Paris he would have been aware of new research and discoveries; and from Buffon's theory of the Earth he would have known of the controversies surrounding the age and development of the world. His own observations would have confirmed or disputed the prevailing orthodoxy and stimulated fresh lines of thought.

He was also keen to learn from others at first hand. In 1784 Barthélemy Faujas de Saint-Fond (1741–1819) made two visits to Edinburgh during a tour of England and Scotland, during which he retraced the steps of Joseph Banks to Fingal's Cave on the island of Staffa. As a young man Faujas had abandoned a career in law and politics to follow his passion for geology and natural history. He had come to the attention of Buffon, who had invited him to Paris and helped to find him employment. He was an expert on mining and had published an important study of the extinct volcanoes of the Vivarais and Velay regions of south-east France. His tour of Scotland was controversial because he identified many geological features as volcanic, a proposition flatly rejected by many local authorities, including John Walker, the professor of natural history at Edinburgh University. Others in the city welcomed him warmly. Joseph Black showed him a chemical kiln he had invented and gave him a sample of petrified wood from Lough Neagh in Northern Ireland. John Aitken, professor of anatomy, showed him a forceps device for delivering babies and a single-barrelled rifle that could fire two shots. (Faujas was impressed by the first, but not by the second.) William Cullen persuaded him that a glass of punch made from rum, lime juice, a little nutmeg and boiling water was the ideal counter to the depressing Scottish weather. Adam Smith astounded him by taking him to a bagpipe competition.

Hutton showed Faujas his mineral samples, but the Frenchman, mistaking the purpose of the collection, was not impressed.

> Doctor James Hutton is perhaps the only private individual in Edinburgh who has brought together in a cabinet a few minerals and a large series of agates, drawn particularly from Scotland; but I found that he was not attached enough to collecting the various matrixes in which they are enclosed, and which serve to complement the natural history of these stones. I therefore had much more pleasure in conversing with this modest scientist, whom I went to see, than to examine his collection, which presented me nothing new in this genre; since I had just observed and studied in place and on nature itself, most of the objects in this collection.[3]

Hutton's collection did include many agates, but there was a lot else besides. The items within it – fossil woods, ores, mineral veins, rock salt, septarian nodules (mostly carbonate concretions but with a distinctive surface pattern (septa) formed during their development) and marbles, among others – were assembled not as part of some geographic geological survey, but for the processes they suggested which had brought about the present state of the Earth's surface and what might have caused them. Hutton's interest was not spatial and static, but dynamic. As the subsequent publication of his theory showed, he was constantly making deductions from the appearance of his samples, in particular his contention that subterranean heat was necessary for the consolidation rock was based on a descriptive analysis of his specimens.[4]

In his seminal book *James Hutton and the History of Geology*, Dennis R. Dean surmises that Hutton would have taken Faujas to look at Arthur's Seat and Salisbury Crags, the rocky outcrops behind his home in St John's Hill. There they would have disagreed about aspects of the origin of the three hills in the centre of Edinburgh – Arthur's Seat, Carlton Hill and the castle rock – which Faujas believed were all volcanic. He claimed: 'the operation of subterraneous fire is manifest everywhere around Edinburgh'.[5] Such a visit is possible – even likely, since Hutton knew the area well and had studied it in detail – but there is no evidence the two men ever made it or argued over the origins of the features. Hutton maintained that there was a difference between lava and basalt (called whinstone in Scotland at the time and present in Salisbury Crags and in Fingal's Cave).

> While Hutton clearly favoured observations supporting the igneous [volcanic] origin of basalt, he demurred from identifying basalt with lava. Though both flowed from volcanoes, lava, he believed, had been erupted from the volcano's mouth or some other surface aperture. Basalt, on the other hand, was wholly subterraneous, though it could be exposed subsequently through the erosion of overlying sediments.[6]

If Hutton and Faujas did disagree, expert opinion would now give them both some support. 'Modern geologists would say that Faujas and Hutton were both partly right: while Arthur's Seat is the remnant of an ancient volcano, Salisbury Crags [next to it] is the glaciated remains of a Carboniferous sill.'[7] Playfair notes that early sketches of Hutton's ideas were found among his papers after his death (although now lost). By the time of Faujas' visit he was nearing a more complete version. Faujas mentions that Hutton was working on his Theory of the Earth 'in the calm of his cabinet'. Whether Hutton discussed it with him, or, since Faujas' book was published after Hutton made his ideas public, he subsequently read it, we do not know. Playfair does not mention Faujas and says that Hutton did not communicate it with anyone except Joseph Black and his fellow anatomy student John Clerk, younger brother of George Clerk Maxwell – now known since his purchase of an estate as John Clerk of Eldin.

Shortly afterwards Hutton was visited by another volcanologist, Sir William Hamilton, who had made extensive studies of Etna and Vesuvius and published his findings in 1772. This time we know (from Hutton's later publication of his theory) that he did take Hamilton to look at Salisbury Crags, to an area now known as the Hutton Section. There Hutton showed him veins of minerals and spars in the whinstone (basalt) and Hamilton confirmed that he had never seen similar occurrences in lava.

In working on the final drafts of his ideas, Hutton was part of a long tradition of publications by scientists and travellers suggesting all or part of a unified theory of how the Earth came to be the way it is. Throughout the 18th century the biblical explanation of the creation and development of the Earth was being challenged, either implicitly, or explicitly. In the previous century most religious scholars agreed on a date for the beginning of the world around 4000 BC, calculated by references in the Old Testament. In 1644 the Cambridge theologian John Lightfoot put the date of Creation at 26 October 3926 BC. Archbishop Ussher, Primate of All Ireland, put the date at 22 October 4004 BC in his *Chronology*, published in 1650. He dated Noah's Flood at 2349 BC, an important event because the increasing realisation that many rocks formed from sedimentary layers at the bottom of the sea could be tied to the biblical narrative. But as the world began to be explored and Europeans began to visit and learn about other cultures, biblical orthodoxy began to give way to a more sceptical attitude. Chinese written histories older than Noah contained no suggestion of a deluge which could have engulfed the Earth and brought about geological changes.[8]

Using St Peter's assertion that a day to God is like a thousand years to man, a more allegorical interpretation of the biblical Six days of Creation became possible and some scientists began to fit their theories into this looser religious framework. Yet as the scientific evidence built up proving that natural processes could not have transformed the landscape so quickly, even this approach became inadequate. Slowly it gave way to deism, the concept that God was the instigator of Creation, but he left it to nature to provide the means and the timetable.[9] Some religious scholars were tolerant of this approach, but even as late as Hutton's time it paid to be cautious about flatly contradicting literal interpretations of the Bible. Buffon, now a fellow of the French Academy of Sciences as well as director of Le Jardin du Roi, had been censured for the first volume of his

Histoire Naturelle of 1749, and his theory in *Epoques de la Nature*, 1778, which estimated that Creation required about 85,000 years. Even this was a toning down of his real views: his notebooks suggest he really believed the Earth was a lot older, perhaps three million years, but was reluctant to say so in public.

It is inconceivable that Hutton had not read any of Buffon's works. From his response to de Luc's criticisms it appears that he could read French, but in any case selections from Buffon's vast series on the natural world were translated into English by William Smellie[10] and published in Edinburgh by William Creech between 1780 and 1785. First was his theory of the Earth, which not only contained his own ideas, but reviews of other contributions by English, French and German writers.[11] Buffon differed from Hutton in believing that an all-engulfing ocean shaped the Earth, then receded leaving the elevated continents, but some of his ideas foreshadowed the concepts taken up by Hutton. These included the erosion of the mountains and land by wind and rain and what we now call 'uniformity', the idea that the processes which shaped the Earth are the same now as they have always been.

> To give consistency to our ideas, we must take the earth as it is, examine its different parts with minuteness, and, by induction, judge of the future, from what at present exists. We ought not to be affected by causes which seldom act, and whose action is always sudden and violent. These have no place in the ordinary course of nature. But operations uniformly repeated, motions which succeed one another without interruption, are the causes which alone ought to be the foundation of our reasoning.[12]

Other theories already published included *An Inquiry into the original state and formation of the Earth*, by John Whitehurst in 1778. He was a Derby clockmaker and amateur geologist and had been a prominent member of the Lunar Society. Hutton may have met him during his time in Birmingham and Derbyshire, but in 1774 Whitehurst had moved to London to take up a post at the Royal Mint. We know that Hutton read his treatise, or a later revision of it, because he cited it in his own theory. Drawing on his own extensive fieldwork in Derbyshire and on his reading of the theories of others, Whitehurst declared:

> It is not my intention to point out the fault in other systems, but to avail myself of such parts of them as are applicable to my own design, namely to trace appearances in nature from causes truly existent; and to inquire after those laws by which the Creator chose to form the world, not those by which he might have formed it had he so pleased.[13]

This was very near what Hutton was intending to do, to draw on the discoveries of others and to deduce from the state of the Earth as it is now the processes by which it had been formed in the past. God was acknowledged as the instigator, but nature provided the means.

Whitehurst's theory combined the idea of a universal primeval ocean with the effects of subterraneous heat and, like Hutton, claimed that the Earth remained both dynamically active and subject to the regular, gradual and continuous processes of nature. However, his scope was much broader than Hutton's, including discussions of the laws of gravity, the shape of the Earth and its rotation around the sun. He departed from Hutton's thinking by trying to fit his narrative into a quasi-biblical framework, suggesting that the original state of the world was chaos and that a massive 'earthquake-cum-volcano' had precipitated the Noah's Flood.[14] He made no attempt to calculate the age of the Earth, but asserted 'the earth had a beginning, and has not existed from eternity, as some people have imagined'. The *Inquiry* was a curious mixture of imagination and observation. Whitehurst supplemented his flights of fancy with some original fieldwork in Derbyshire and Staffordshire, Wales and Northern Ireland, where at the request of the Royal Society in London, he had visited the Giant's Causeway, a spectacular display of hexagonal basalt columns similar to Fingal's Cave.

Whitehurst's interest was understandable. He had a record of studying the geology of his native county and he may have rehearsed his theories in front of his friends and been encouraged to publish them – although we cannot be sure because the Lunar Society kept no minutes of its meetings. The publication in 1780 of another book, *The Antiquity and Duration of the World*, appeared to come from nowhere. Its author was George Hoggart Toulmin, a young doctor with a practice in Wolverhampton. He had studied medicine and graduated from Edinburgh University the year before and had hitherto shown no apparent interest in geology. Nor had he done any fieldwork. Nevertheless, his book (which was revised and reprinted under different titles in 1783, 1785, and 1789) made many of the same points Hutton was to make in his *Theory of the Earth*.

Rather than couching his rejection of religious calculations of the Earth's age in vague terms, or hiding behind a pseudonym as others had done, Toulmin was contemptuous of earlier attempts to estab-lish a chronology of the Earth's history based on the Old Testament. He accepted the Aristotelian belief in the eternity of the world and claimed that the organic and inorganic matter of which the Earth is composed was in a state of constant flux, resulting from decay and erosion. New fossiliferous sediments were being deposited in the oceans constantly, mountains were being destroyed by erosion and new ones would be formed by elevation. The operations of nature proceeded in a slow and uniform way, each part of the universe operating in a manner designed to secure the preservation of both the parts and the whole.[15]

Many of Toulmin's ideas were so similar to Hutton's and the phrases he used so alike those contained in Hutton's theory when it was eventually published, that some scholars believe either that Hutton had read Toulmin's book before he wrote the final version of his theory, or vice versa, that Toulmin had somehow obtained an early draft of Hutton's work.[16] Whether a written draft of Hutton's theory existed in 1776–9 when Toulmin was a student we don't know – if it did, was it at a sufficiently advanced stage for the med-ical student to have plagiarised the language? It is very possible that Toulmin attended Joseph Black's chemistry lectures in which he discussed some geological concepts, particularly the effect of heat on certain minerals, such as limestone, which he had tested in experiments with Hutton.

It is also possible that both men arrived at their ideas independ-ently. Although there are similarities in their theses, there are also important differences. Toulmin gave God no place, whereas Hutton was careful to cite God as the instigator. Toulmin's book dealt with the place of Man within the system, Hutton ignored Man. Toulmin thought the world was eternal. Hutton, although he conceded it was impossible to give a date for a beginning or for an end, made no claim about the eternity of the Earth. A very important difference was that Toulmin's conclusions were based on no fieldwork; instead he had read very widely and synthesised the opinions of others into a coherent theory of his own. This was one of the reasons his books were largely ignored by scientific opinion of the day, but there were others, outlined by the historian Roy Porter:

> Toulmin ... dealt head-on with an issue – the natural history of Man – which was so fraught with tension that practically the whole community of naturalists investigating the history of the Earth chose to set it to one side.

Whereas Hutton, although he did not speculate on the history of Man, gave him his place:

> Hutton might – like Toulmin – discard Scripture and scorn the Noachian Deluge as a geological cause. He might assert that in the system of the Earth there was no vestige of a beginning and no prospect of an end. But his published *Theory* implied that Man – uniquely in the geo-system – was not a creature subject to endless cycles of dissolution and transformation; and here Hutton and Toulmin totally part company.[17]

In Hutton's vision: 'The globe of this earth is evidently made for man. He alone, of all the beings which have life upon this body, enjoys the whole and every part; he alone is capable of knowing the nature of this world, which he thus possesses in virtue of his proper right; and he alone can make the knowledge of this system a source of pleasure and the means of happiness.' The Earth had been constructed as a habitat for Man, who completed the Creation because he alone understood it.[18]

But if they excited little interest or comment on scientific grounds, Toulmin's books provoked outrage among churchmen and those on the right of politics. They were viewed as a deliberate challenge to Christian belief. John Ogilvie, in his *An Inquiry into the Causes of the Infidelity, and Scepticism of the Times* (1783), branded Toulmin a 'free thinker' alongside David Hume and the historian and Whig politician Edward Gibbon. The Rev. Ralph Sneyd, in his *A Letter to Dr Toulmin M.D. Relative to his book on the Antiquity of the World* (1783), saw Toulmin as conspiring to destroy morality, justice, and society, demanding that although 'the rack and gibbet' were now obsolete for punishing blasphemers, it was a disgrace that Toulmin had been allowed to publish and advocated that he be imprisoned.[19] Even John Robison, professor of physics at Edinburgh University, and friend of Hutton and Black, joined in the attack in his pamphlet *Proofs of a Conspiracy against all the Religions and Governments of Europe* of 1798.

There were lessons here which Hutton was careful to learn, but that could not wholly protect him from attack.

Chapter 14

Time . . . is to nature endless

JAMES HUTTON WAS due to make the first public presentation of his theory to a meeting of the Physical Class of the Royal Society of Edinburgh on 7 March 1785. The meeting was to be chaired by William Cullen, professor of medicine and one of the four presidents of the Physical Class. The Society had no premises of its own, so met in the library of the university, which was still housed in a ramshackle collection of inadequate buildings to the south of the Old Town at Kirk o' Field. Of the two divisions, the Physical Class was the more vigorous and its meetings were better attended than those of the Literary Class. Hutton had attended one literary meeting in 1789 in the company of Adam Smith, the writer Henry Mackenzie and the poet Samuel Rogers. The four of them made up more than half the audience and Smith fell asleep during the long and tedious presentation on the law of debt.

But although he might be reasonably confident of a larger and more attentive audience, Hutton could not count on a sympathetic one. Geology – and in particular geological theorising – was not yet a particular interest of the RSE fellows. Many were sceptical of its value, particularly in light of its recent history with the reception given to Whitehurst and Toulmin. They would also be alert to the religious implications of any new thesis. Several of the leading members of the society were ordained ministers in the Church of Scotland, including William Robertson, the university principal, John Playfair, appointed professor of mathematics in that year, and John Walker, professor of natural history. Walker, one of the secretaries of the Physical Class was known to disparage any scientific theories that could not be tested by experiment and disdained those who proposed theories of the Earth – calling them 'system mongers'.[1]

When the day came Hutton was ill and unable to appear. He had presented controversial papers previously, such as his 'Theory of Rain', so there is no reason to believe his indisposition was feigned. It could possibly have been an early sign of the bladder stones which were to cause him considerable pain later. In his place the paper was read by his close friend Joseph Black. Its full title was 'Theory of the Earth; or an investigation of the laws observable in the composition, dissolution and restoration of land upon the globe'. Black's fluency and clear explanations had made his chemistry lectures very popular; now he used his oratorical skills to present Hutton's ideas in as simple and straightforward way as he could. He started with a clear declaration of religious faith. Hutton was not going to fall into the trap that had snared Toulmin by denying or ignoring God. He may not believe in the six days of Creation or Noah's Flood, but he asserted from the start his acceptance of a divine inspiration behind the process he was about to describe. 'We perceive a fabric, created in wisdom, to obtain a purpose worthy of the power that is apparent in the production of it.' And later: 'more obvious . . . are the presence and efficacy of design and intelligence in the power that conducts the work.'[2]

The purpose and design of the world were to provide a habitat for living things, including Man. Although the inspiration may be divine – and therefore beyond question – Hutton asserted the right of science to 'find a fit subject of investigation in every particular, whether of form, quality or active power that presents itself in this system of motion and of life'.

> If, in pursuing this object, we employ our skill in research, not in forming vain conjectures, and if data are to be found, on which science may form just conclusions, we should not long remain in ignorance with respect to the natural history of this earth, a subject on which hitherto opinion only, and not evidence, has decided.

He pictured the Earth as a machine, with three constituent parts – earth, sea and air – and with forces acting on them – gravity, light, heat, cold and condensation. Then he described the action of the machine: soil, necessary for plants to grow, was created from the destruction of rock, but soil was itself destroyed by the action of the weather and washed into the sea.

> The surface of this land, inhabited by man, and covered with plants and animals, is made by nature to decay, in dissolving from that hard and compact state in which it is found below the soil; and this soil is necessarily washed away, by the continual circulation of the water, running from the summits of the mountains towards the general receptacle of that fluid.
>
> The heights of our land are thus levelled with the shores; our fertile plains are formed from the ruins of the mountains, and those travelling materials are still pursued by the moving water and propelled along the inclined surface of the Earth. These moveable materials, delivered into the sea, cannot, for a long continuance, rest upon the shore; for, by the agitation of the winds, the tides and currents, every moveable thing is carried farther and farther along the shelving bottom of the sea, towards the unfathomable regions of the ocean.

If this were the end of the process, the machine must inevitably wind down. But then Hutton produced a controversial concept – infinite time. The idea that time had no beginning and no end dated at least from Aristotle's *Physics*, but was rejected by Christian, Jewish and Islamic philosophers, who held that time began when God created the world. To Hutton, this made no sense.

> Time, which measures everything in our idea, and is often deficient to our schemes, is to nature endless and as nothing; it cannot limit that by which alone it had existence; and as the natural course of time, which to us seems infinite, cannot be bounded by any operation that may have an end, the progress of things upon this globe, that is, the course of nature, cannot be limited by time, which must proceed in a continual succession.

But if time was infinite and the process of destruction had no limit, then the machine must be capable of repairing itself. This was the

In later life, Hutton became an Edinburgh celebrity and was one of the earliest subjects for the portrait painter Henry Raeburn. *Album/Alamy Stock Photo*

Above. Hutton's father William was a prominent merchant and City Treasurer but died when Hutton was three years old. *The Merchant Company of Edinburgh*

Right. Hutton's name on the matriculation register at Edinburgh University, then known as the 'Tounis College'. John Davie, his friend and later business partner, signed the same day. *University of Edinburgh*

Above. Alexander Monro's anatomy dissections were highly popular, but a shortage of corpses was a constant problem. Portrait by Henry Raeburn. *National Galleries of Scotland. Purchased 1990*

Left. John Clerk of Eldin was Hutton's friend since university and a later collaborator in his geological expeditions. *The History Collection/Alamy Stock Photo*

Above. G. F. Rouelle gave
Hutton his first lessons in
geology as part of his
chemistry course in Paris,
enlivening his lectures with
demonstrations and
experiments. *Sheila
Terry/Science Photo Library*

Right. Le Comte de Buffon's
Natural History of the Earth
challenged the Biblical
narrative of the Creation
but was careful to give God
a place in evolution.
Wellcome Collection

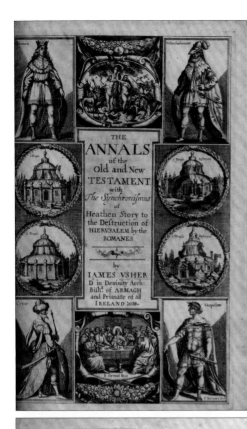

THE
ANNALS
OF THE
WORLD.

Deduced from
The Origin of Time, and continued to the
beginning of the Emperour *Vespasians* Reign, and the
totall Destruction and Abolition of the Temple
and Common-wealth of the *Jews.*

Containing the

HISTORIE
Of the OLD and NEW
TESTAMENT,
With that of the
MACCHABEES.

Also all the most Memorable Affairs of *Asia* and *Egypt*,
And the Rise of the Empire of the *Roman Cæsars*,
under *C. Julius*, and *Octavianus.*

COLLECTED
From all History, as well Sacred, as Prophane, and Methodically digested,

By the most Reverend *JAMES USSHER*, Arch-
Bishop of ARMAGH, and Primate of IRELAND.

LONDON,
Printed by E. TYLER, for J. CROOK, at the Sign of the
Ship in St. Pauls Church-yard, and for G. BEDELL,
at the *Middle-Temple-Gate*, in *Fleet-Street*. M. DC. LVIII.

A 2

Above. Archbishop Ussher calculated that
the world had been created on 22 October
4004 BC and also dated Noah's flood.

Left. Hutton presented his dissertation to
the University of Leiden and received the
degree of Doctor of Medicine. *University of
Leiden, Bert Schuchmann*

DISSERTATIO PHYSICO-MEDICA
INAUGURALIS
DE
SANGUINE ET CIRCULATIONE
MICROCOSMI.
QUAM,
ANNUENTE DEO TER OPT. MAX.

Ex Auctoritate MAGNIFICI RECTORIS,

D. JOACHIMI SCHWARTZII,
JURIS UTRIUSQUE DOCTORIS ET ANTECESSORIS
IN ACADEMIA LUGDUNO-BATAVA.

NEC NON

Amplissimi SENATUS ACADEMICI *Consensu,*
& Nobilissimae FACULTATIS MEDICÆ *Decreto,*

PRO GRADU DOCTORATUS,

Summisque in MEDICINA Honoribus & Privilegiis rite ac
legitime consequendis,

Eruditorum disquisitioni submittit

JACOBUS HUTTON, SCOTO BRITTANNUS.

Ad diem 12. Septembr. 1749. H. L. Q. S.

LUGDUNI BATAVORUM,
Apud WILHELMUM BOOT, 1749.

Above. Hutton introduced the lighter, faster Norfolk plough to his farm in Berwickshire, replacing the heavy Scottish wooden implement. *Evelyn Simak/Wikimedia Commons*

Right. Sal ammoniac production from soot in Egypt – Hutton and Davie used a similar process in Edinburgh.

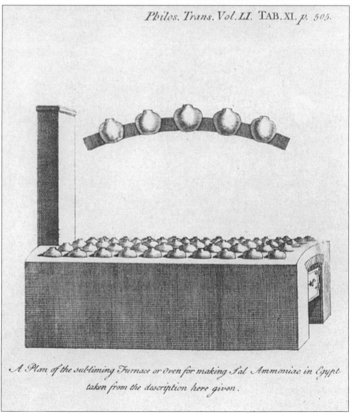

Philos. Trans. Vol. LI. TAB. XI. p. 505.

A Plan of the subliming Furnace or Oven for making Sal Ammoniac in Egypt taken from the description here given.

The chemist and doctor Joseph Black (right) became Hutton's closest friend and supporter. Engraving from John Kay's *A series of original portraits and caricature etchings . . . 1837–38*.

Hutton may have met James Watt while working on the Forth & Clyde canal. They became firm friends. *Wellcome Collection*

Adam Smith, author of the *Wealth of Nations*, founded the Oyster Club with Hutton and Black.
Engraving from John Kay's *A series of original portraits and caricature etchings . . . 1837–38.*

Hutton organised and catalogued the fossil collection of Princess Dashkova during her stay in Edinburgh while her son attended the university. *Pictorial Press Ltd / Alamy Stock Photo*

John Playfair, kirk minister and mathematician, was Hutton's first biographer and tried to popularise his work. *Wellcome Collection*

Among Hutton's fiercest critics was the Irish scientist and geologist Richard Kirwan, who accused him of atheism. *Wellcome Collection*

Jean-André de Luc attacked Hutton's explanation of rain and his *Theory of the Earth*. *Wellcome Collection*

Right. James Tassie depicted Hutton as a philosopher in a toga. *Wellcome Collection*

Below. Siccar Point: Hutton could read the cliff like a history of the earth, formed over countless eons of time. *Angus Miller, Geowalks*

crux of his argument and he recognised the huge leap of imagination he was asking his audience to make:

> Animated with this great, this interesting view, let us strictly examine our principles, in order to avoid fallacy in our reasoning; and let us endeavour to support our attention, in developing a subject that is vast in its extent, as well as intricate in the relation of parts to be stated.

Against the magnitude of infinite time, the history of Man was insignificant. The history of animals, particularly those which originated in the oceans, was older and the fossil record gave us some means of calculating how long these natural processes took to occur. 'But how shall we describe a process which nobody has seen performed, and of which no written history gives any account?' His answer was that from examining what was happening to our world now, we could deduce what happened in the past and predict what would happen in the future. This concept, now called 'uniformity',[3] had already been suggested by Buffon. Hutton writes:

> In examining things present, we have data from which to reason with regard to what has been; and, from what has actually been, we have data for concluding with regard to that which is to happen hereafter. Therefore, upon the supposition that the operations of nature are equable and steady, we find, in natural appearances, means for concluding a certain portion of time to have necessarily elapsed, in the production of those events of which we see the effects.

He then starts to examine the data: fossil shells embedded in marble or limestone at the tops of mountains were once the bodies of creatures which, after their deaths, were compacted into strata at the bottom of the oceans. This process accounted for 90 or perhaps even 99 per cent of the Earth's surface that we can see. An exception to this was granite, which had a different origin (which Hutton was to deal with later). But this prompted questions: how were these individual creatures formed into a solid mass; and how was this mass raised up to become the dry land on which we live? It could not be explained by any action of the sea, or of the forces produced by the Earth's rotation or movement in the solar system. To find an answer, we had to look to the science of chemistry.

At this point, the chemist Joseph Black who had been reading Hutton's words, ended his presentation. When the society met again on 4 April Hutton was sufficiently recovered to read the remaining parts of his paper himself. The first part of the theory, although it made demands of his learned audience, was reasonably succinctly stated. Perhaps Black, the experienced and accomplished lecturer, had modified Hutton's written brief, eliminating unnecessary digressions to concentrate on the essential elements of the theory? Hutton, who Playfair describes as clear, simple and concise in his conversation, but the opposite in his written communications, appears to have stuck to the lecture he had written out. If his audience was hoping for a quick answer to the questions his final remarks in the last session had posed, they were disappointed.

He firstly treated them to a discussion of the ways in which the mass, before it could become solidified, could be made fluid – by the action of heat, or by a solvent, he suggested – by fire or by water. After what Playfair, who was almost certainly in the audience, describes as a 'long and minute examination', Hutton came to the conclusion that water could not have been the agent producing the fluidity, it must have been subterranean heat. But then, rather than going on to explain how continents had been formed, he went into a long digression on the nature of strata – describing the 'siliceous', the 'sulphurous' and the 'oily or bituminous'. To illustrate this part of his talk he gave examples from his travels around Britain, from the Isle of Wight to the Cheshire salt mines, to Portsoy in the northeast of Scotland. He also drew at length on the mineral samples in his collection and from his voracious reading. He cited discoveries from Spain and Germany, a Hungarian mine and a Siberian iron ore deposit.

At one stage he admitted: 'It would be endless to give examples of particular facts, so many are the different natural appearances that occur, attended with a variety of different circumstances' before continuing to cite even more. He also claimed that 'phlogiston' was present in some of the minerals he was describing. The concept of 'phlogiston' as a substance necessary for combustion had long been challenged by the French chemist Antoine Lavoisier, who showed that oxygen, not the mythical phlogiston, was the vital factor. Even Black, Hutton's closest friend and collaborator, who had taught the phlogiston theory for 15 years, had abandoned it by the 1780s,[4] but Hutton clung to it. In this long digression he demonstrated his extraordinary knowledge of geology, based on his own observations and experiments and wide reading of the literature. Buried in his explanations were indications of the proofs he would later produce in answer to his critics, but in the mass of verbiage the significance was possibly lost on most of his audience.

Hutton finally answered his own question on how continents were formed in part III of his lecture. After considering alternatives, he concluded that the only power capable of raising the compacted rock at the bottom of the seas above the level of the water to form not only dry land, but mountain ranges such as the Alps or the Andes, was subterranean heat. This made the mass expand and forced it upwards. His proof for this was that strata, which were laid down horizontally on the seabed, could be found at very different angles:

The strata of the globe are actually found in every possible position: for from horizontal they are frequently found vertical; from continuous they are broken and separated in every possible direction; and from a plane they are bent and doubled. It is impossible that they could have originally been formed by the known laws of nature in their present state and position.

Could any other force have been responsible for this? No, he concluded: 'This agent is matter actuated by extreme heat and expanded with amazing force.' The evidence was before us:

We have but to open our eyes to be convinced of this truth. Look into the sources of our mineral treasures, ask the miner from whence has come the metal into his vein? Not from the earth or air above, not from the strata which the vein traverses; these do not contain one atom of the minerals now considered: there is but one place from whence these minerals may have come; this is, the bowels of the earth, the place of power and expansion, the place from whence must have proceeded that intense heat by which loose materials have been consolidated into rocks, as well as that enormous force by which the regular strata have been broken and displaced.

This was a crucial part of Hutton's argument. The proof that the force which created the world as we knew it was heat from below was to be seen in veins of minerals injected, as it were, into strata. These introduced minerals were quite different from the surrounding rocks into which they had been forced; they were not of the kind found in the strata laid down in the seas, they must have come from deeper in the Earth, from 'the unknown region, that place of power and energy'. This concept of one mineral being injected into another was not original. Hutton had read it in the works of the Swedish chemist and mining expert Axel Fredrik Cronstedt, whose book on mineralogy had been translated into English in 1772.[5]

Hutton went further, saying that the huge disruptive forces which could shatter and upend layers of rock could happen time and again in series. The evidence was there to be seen – and here he defended his own method of inquiry:

> It is very common to see three successive series of those operations; and all this may be perceived in a small fragment of stone, which a man of science may examine in his closet, often better than descending to the mine, where all the examples are found on an enlarged scale.

How marble, for example, a mineral formed at the bottom of the sea came to be found on the tops of mountains was not immediately obvious to many. Yet Sicilian Jasper, a type of marble which Hutton had in his collection, had been found high on Mount Etna, a volcano capable of spewing molten rock miles into the air. Who, after knowing this, could doubt that Sicily itself had been raised from the ocean bed by such a force?

> When fire bursts forth from the bottom of the sea, and when the land is heaved up and down, so as to demolish cities in an

instant, and split asunder rocks and solid mountains, there is nobody but must see in this a power, which may be sufficient to accomplish every view of nature in erecting land, as it is situated in the place most advantageous for that purpose.

In the last part of his presentation, Hutton showed how this raised rock could be eroded by sun and rain and the resulting debris combined with vegetable matter to form soil, which was necessary for plants to grow and in turn support animal life. But this soil was itself being eroded, washed to the sea by the action of wind, rain and rivers. This was a gradual process, but whole continents could be destroyed in this way:

> The destruction of the land is an idea that does not easily enter into the mind of man in its totality, although he is daily witness to part of the operation. We never see a river in a flood, but we must acknowledge the carrying away of part of our land, to be sunk at the bottom of the sea; we never see a storm upon the coast, but we are informed of a hostile attack of the sea upon our country; attacks which must, in time, wear away the bulwarks of our soil, and sap the foundations of our dwellings. Thus, great things are not understood without the analysing of many operations, and the combination of time with many events happening in succession.

Had Hutton stopped there he may have provoked a reaction from his audience. His theory was revolutionary. He had presented the idea of the Earth as a machine revolving in a constant cycle of destruction and renewal. Many common types of rock had been formed from the bodies of sea creatures deposited on the seabed and consolidated into dense masses by compression from the weight of water above them and – he claimed – by the action of subterraneous heat. Enormous powers, deep in the Earth – a region, he admitted, about which we know very little – shattered the strata and pushed rock upwards above the level of the water. The mighty mountains and continents raised in this way were slowly ground down and washed into the sea – to be deposited on the seabed for the process to begin over again. He had introduced difficult concepts – such as infinite time – had allowed God as the instigator, but had given the deity no part in the working of the machine.

But in his effort to convince he had buried his simple yet challenging concepts in a morass of evidence. He listed and described gravels, sands and clays, showed the difference between fossilised woods and mineralised woods, sought to measure the slow wearing down of the land and the retreat of the seas by reference to ancient Greek and Roman sources, and at each stage defended the logic of his conclusions by considering and dismissing the alternatives. He left his (by now surely exhausted) audience with a conclusion and the ringing phrase which has come to characterise his theory:

> We have now got to the end of our reasoning; we have no data further to conclude immediately from that which actually is, but we have got enough; we have the satisfaction to find, that in nature there is wisdom, system, and consistency. For having, in the natural history of this earth, seen a succession of worlds, we may from this conclude that there is a system in nature; in like manner as, from seeing revolutions of the planets, it is concluded, that there is a system by which they are intended to continue those revolutions. But if the succession of worlds is established in the system of nature, it is in vain to look for anything higher in the origin of the earth. The result, therefore, of our present enquiry is, that we find no vestige of a beginning, no prospect of an end.

A theory which Playfair was able to summarise in three or four pages of his biography took Hutton 95 pages to expound.* To his disappointment, the theory provoked neither immediate opposition nor support – it fell flat. Playfair commented:

> It might have been expected, when a work of so much originality as this *Theory of the Earth*, was given to the world, a theory which professed to be the result of such an ample and accurate induction, and which opened up so many views, interesting

---

* Both appeared in the same format in the *Transactions of the Royal Society of Edinburgh*.

not to mineralogy alone, but to philosophy in general, that it would have produced a sudden and visible effect, and that men of science would have been everywhere eager to decide concerning its real value. Yet the truth is, that it drew their attention very slowly, so that several years elapsed before anyone shewed himself publicly concerned about it, either as an enemy or a friend.[6]

Chapter 15

His ideas are magnificent . . .

HUTTON WAS CLEARLY disappointed by the lack of reaction to his theory and perhaps, told by his friends that his two-part lecture had been too long, too detailed and too discursive to have been easily absorbed, began working on a shorter and simpler version. His intention appears to have been to have a few copies privately printed so that he could circulate it among friends, supporters and correspondents. He was also keen to head off any religious criticism which might be attracted by his implicit rejection of the Old Testament narrative, his introduction of the concept of infinite time – a direct contradiction of the very finite age of the Earth calculated by the biblical scholars – and his depiction of the Earth as a self-perpetuating machine, rather than as God's creation. He began to sketch out a 'Memorial Justifying the Present Theory of the Earth from the Suspicion of Impiety' which would stand as a preface to his new 'Abstract' and, he hoped, acquit him of any charge of atheism.

His draft of his intended preface ran to 1,000 words and was as opaque and contorted as his original theory. It began with an assertion that the purposes of religion (the 'revelation' of the Bible) and science ('natural philosophy') were so different that there could be no conflict between the two.

> The object of revelation and that of natural philosophy being thus perfectly different, it must be absurd to suppose that these can truly interfere, as this may only happen in supposing them not strictly adhering to their respective subjects and one or other of them as not being just.[1]

He then entered into a lengthy philosophical argument with himself about whether religion or science could be considered the higher authority, before arriving back at his original conclusion that they were so different that they could not be compared. He discussed the Jewish texts on which the Christian religion was based and whether they could be relied upon as guides to the natural history of the Earth. He then unwisely entered into one of the controversies which had engulfed previous scientists who had questioned the literal truth of the Six Days of Creation: 'It would be unreasonable, or no less than absurd, to suppose the term "day," by which each of those periods is expressed in this Jewish history, means anything besides an indefinite period, or means any more than to signify that God made all things in a certain order.' After another impenetrable paragraph he concluded: 'Consequently, although this theory of the earth gives a most distant view with regard to the operations of nature, it does not in any respect interfere with the chronology of the Old Testament.'

If Hutton believed that this 'memorial' would protect him from religious criticism he was being naïve. Second thoughts, or perhaps the urging of his friends, persuaded him to take advice. He sent his draft to William Robertson, principal of the university and the man who did more than any other to bring about the creation of the Royal Society of Edinburgh, which had given Hutton his platform. Besides being a distinguished historian and a highly successful author, Robertson was an ordained minister in the Presbyterian Church and a free thinker who had included the atheist David Hume among his friends. He was also very familiar with just the sort of controversy that Hutton was desperate to avoid. His daughter Mary was married to Patrick Brydone, a traveller and writer who had published a book on his journeys around the islands of Sicily and Malta. In it he described meeting the geologist-priest Giuseppe Recupero, who had excavated a site on the side of Mount Etna and uncovered seven strata, each of which he estimated to represent at least two thousand years. From this he calculated the age of the volcano to be at least 14,000 years and was embarrassed to have contradicted the biblical age of the Earth.[2] This, to us fairly innocuous piece of reporting, led to Brydone's book being criticised by Boswell and Johnson and to a complaint from the philosopher James Beattie to the judge Lord Hailes.

Robertson tactfully rewrote Hutton's text, reducing the length by two-thirds and avoiding inflammatory language. But in the letter which enclosed the new draft, he strongly advised Hutton to drop the idea of a preface altogether.

> Dear Doctor
>
> I send you a sketch of a short advertisement, which maybe prefixed to your dissertation. The ideas are mostly taken from your own papers, only the style is rendered a little more theological. After all, it requires some debate perhaps whether the dissertation may not be printed without anything prefixed. I purposed to have called upon you today in order to have discussed this point but have so many things to do previous to my going out of town on Monday that it has not been in my power. I shall return in three weeks; if you should begin to print before that time, consult our friend Mr Smith, and on following his advice you will be safe. I return your papers and am ever,
>
> Yours most faithfully
> William Robertson
> College July 22[3]

Whether Hutton did consult Adam Smith, we do not know, but he abandoned the idea of prefacing his Abstract with a religious declaration. When the new paper appeared it was not only shorter – at 2,500 words less than a tenth of the length of the original – but had a number of other significant differences. Its title was changed. No longer was it an *Investigation of the laws observable in the composition, dissolution and restoration of land upon the globe,* but a *Dissertation on the System of the Earth, its Duration and Stability.* Its language was also different – more direct and concise, shorn of digressions and most of the illustrations and examples. Its opening paragraph succinctly described Hutton's concept of the Earth as a machine and although it was 'wisely adapted' and 'wisely calculated to fulfil the purpose for which it was designed', there was no direct reference to God as the inspiration behind the machine. As with Hutton's pamphlet on coal and culm, the marked difference in style of the Abstract suggests that his work may have been edited. Perhaps Robertson, an accomplished and successful author, rewrote the paper as well as the intended preface.[4]

The longer version of Hutton's paper, as it had been delivered in March and April 1785, was accepted for publication in the first volume of the *Transactions of the Royal Society of Edinburgh*, along with his 'Theory of Rain', but that would not appear for another three years.[5] In the meantime he had the whole paper published as a pamphlet. It seems that only a few were printed, but they circulated far and wide. Professor Adam Ferguson sent a copy to the Swiss geologist Horace Bénédict de Saussure, whose book *Voyages dans les Alpes* had been read with approval by Hutton, although he did not agree with all its conclusions, particularly over the origin of granite. Hutton had already sent a copy to Saussure at the Academy of Agriculture in Paris, but was so anxious that the Swiss scientist should read it that he provided a duplicate. Ferguson's covering letter praised the *Theory of the Earth*: 'He has long worshipped the same divinities as you and embraced every specimen of stone and earth with the most pious attention. His ideas are magnificent and, what is more precious and more difficult in science, formed with a scrupulous regard for reality.' Ferguson suggests that he and Hutton might visit Saussure in Geneva and see the mountains he had studied for themselves, but the trip never happened.[6]

Hutton also sent a copy to Matthew Guthrie, who had trained as a surgeon in Edinburgh, but was now working as a doctor to the Russian army in St Petersburg. They may have met as members of the Philosophical Society and Guthrie, like Hutton, had become a founder member of the RSE. Guthrie praised the short work but wanted to read the full theory. They also exchanged mineral samples.[7]

Josiah Wedgwood's son brought copies from Edinburgh to Stoke-on-Trent and his father forwarded two to James Watt and Matthew Boulton in Birmingham. Joseph Black sent a summary to Princess Dashkova, who by now had become director of the Russian Imperial Academy of Sciences. She had invited Hutton to lecture in St Petersburg, but through Black he excused himself by saying he was too busy. Black warmly recommended Hutton's ideas:

In the theory of Dr Hutton there is a grandeur & sublimity by
which it far surpasses any that has been offered. The boundless
pre-existence of time and of the operations of Nature which he
brings into our View, the depth & extent to which his imagination has explored the action of fire in the internal parts of the
Earth strike us with astonishment.[8]

Copies of the offprint, or translations, also made their way to
Germany and France, where they were noticed in scientific journals
and found among the papers of the German geologist Abraham
Werner at Freiberg and of Nicolas Desmarest in France, who was
later to include a 50-page discussion of Hutton's ideas in the
Encyclopedie methodique.[9]

However, it was the reception of the paper in Britain that was to
cause Hutton trouble. Following the publication of the *Transactions
of the Royal Society of Edinburgh* in 1788, the theory was summarised and reviewed in several leading journals. The *Analytical Review*
dismissed it in a paragraph as just another geological theory, adding
that these 'usually contain a selection of interesting facts, but their
general systems wanting proof can be considered only as philosophical romances'.[10] The *Critical Review* was fairer, giving four
pages to Hutton's theory and, although it took issue with some of
his assertions, did not condemn him on religious lines.

The *Monthly Review* devoted two pages to Hutton's ideas, but
ultimately fastened on his final phrase 'we can find no vestige of a
beginning, no prospect of an end'. It seems clear that what Hutton
was describing was the process of destruction and renewal of the
land, which was the central concept of his theory. The fact that he
had not found a beginning to this cycle signified that he had not
looked for one; he was not interested in the beginning of the world,
any more than he was in speculating how it might end. What his
critics took him to mean was that there was no beginning or end to
the age of the Earth – that it was eternal and therefore had not been
created by God. The misunderstanding of this ringing phrase –
perhaps wilful in some cases – was to dog Hutton for several years.
He was not helped by the appearance in 1785 of the third edition of
Toulmin's book, which asserted the eternity of the Earth and denied
the role of God. Neither of these controversial claims were in

Hutton's *Theory of the Earth*, but there were many striking similarities in the two theses and even in their phrasing. In 1789 Toulmin published yet another edition, this time under the title *The Eternity of the Universe* – a direct challenge to the biblical scholars – and specifically mentioned Hutton.

A direct attack came from the Scots mineralogist John Williams in his book *The Natural History of the Mineral Kingdom*, published in 1789. After praising Hutton as 'a naturalist of eminent abilities, whose knowledge in several branches of mineralogy does honour to his country', he set out to refute his arguments in abusive terms. For example, on Hutton's assertion that sedimentary rocks had been formed under the ocean and were continually being formed from the earth and stone washed down from the land, Williams claimed that he had 'fully answered and refuted before it was written at least before it was published in my examination of the system of Buffon in his *Theory of the Earth*'. He added: 'I will venture to say the candid intelligent naturalist will [agree] with me, that I have not left the Doctor [Hutton] so much as a particle of earthy matter to form one of his future worlds.'[11]

But his most devastating charge was that Hutton had denied the place of God in Creation and he went on to throw back Hutton's own words, again quoting the infamous phrase of no beginning and no end. 'That Dr Hutton aims at establishing the belief of the eternity of the world is evident from the whole drift of his system and from his own words for he concludes his singular theory with these singular expressions.' Williams concluded:

Thus our modern philosophers labour hard to confirm their favourite scepticism ... they labour hard to rob us of our best inheritance both here and hereafter, to sap the foundations of our belief in revelation and of the superintending care and love and of the over ruling providence of the all benevolent all powerful God our Saviour, who cares for us and upholds us through all the stages of our existence, and like actual robbers these philosophers give as nothing in exchange for our natural inheritance.

If they say that we are poor mistaken ignorants and that they wish to convince us of our error – this is worse than nothing. If we err in charity let us live and die in this error. It is more happy to live in a full persuasion in a feeling sense of the love of God and man while here and in the confident hope of eternal felicity hereafter than to suppose that there is no such thing – that these divine faculties and propensities of our souls which make us capable of loving God and man of admiring God in his works and of ranging thro his creation with sublime delight shall perish for ever and sink into the horrible gulf of non-entity.

Let us turn our eyes from the horrid abyss and stretch out our hands and cry Save [us] Lord or we perish.[12]

The *Monthly Review,* the most persistent of Hutton's critics, devoted 11 pages to Williams' book, the first six on his attack on Hutton.[13] Between 1790 and 1791 it published four letters by Hutton's old adversary Jean André de Luc, who had criticised his 'Theory of Rain'. De Luc's criticisms were geological rather than religious and Hutton attempted to answer them in a letter to the *Monthly Review* which the editor refused to print. He had to wait until his own book was published four years later to reply in public and refute de Luc's claims.[14]

The religious backlash against science and theories of the Earth which did not accept the biblical narrative was being played out in a rising political temperature. In 1789 the Storming of the Bastille and the start of the French Revolution began to polarise political and religious ideals in Britain, typified on the right by the publication of Edmund Burke's *Reflections on the Revolution in France* in 1790 and on the left by Thomas Paine's *Rights of Man* the following year. Both sold in their tens of thousands. Paine's ridiculing of Genesis as a historical narrative (in his later book *The Age of Reason*) equated atheism with revolutionary politics in the minds of many.

De Luc became a prominent proponent of the establishment pro-biblical view, although unwittingly. Beginning in 1791 he had written seven long letters to the Hanoverian geologist Johann Friedrich Blumenbach in which he outlined a theory of the Earth consistent with the account of Moses, as written in the book of Genesis. The fact that there were fossils at the top of mountains merely showed that the seas once covered the tops of mountains. Strata which had been broken and deformed did not indicate the great antiquity of the Earth any more than the famous ruins of Palmyra, where huge stones were piled on top of each other in disarray. But that city had been built and destroyed within the historical period. This 'fact' he claimed, 'at one blow destroys all the systems of geology in which slow causes, acting through an innumerable sequence of ages, were used to explain their formation'.

He directly criticised Hutton's belief that sedimentary rocks had been hardened by fusion as 'reviving an hypothesis which all attentive naturalists had abandoned'.[15] Blumenbach arranged for de Luc's letters to be translated from French into German and published in a scientific journal. However, they were retranslated into English, possibly without de Luc's knowledge or consent, and published in the *British Critic*, a conservative and high-church review designed to counter revolutionary ideas. The publication used – and perhaps distorted – de Luc's views to press its own agenda, declaring that they provided 'demonstrative evidence against those who delight to calculate a false antiquity to the world inconsistent with the sacred records'.[16]

Hutton, as a scientist with no published views on politics, was not immune from the growing tide of opinion against radical views, fanned by events in France, particularly the execution of Louis XVI in 1793 and Robespierre's Reign of Terror, begun later that year. The politician Henry Dundas, member of parliament for Edinburgh, home secretary and effectively minister for war in the government of William Pitt, had cemented Tory hegemony in Scotland. His nephew Robert Dundas, lord advocate – the senior law officer and government minister north of the border – prosecuted a group of leading radicals, including the lawyer Thomas Muir, who was convicted of sedition and transported to Australia. The anti-religious policies of the revolutionaries in Paris encouraged the identification of any opinions contrary to established religious orthodoxy with radical politics. Against this background a charge of atheism was damaging and could be dangerous.

In February 1793 another of Hutton's former detractors returned to the attack. Richard Kirwan had returned to Ireland and was now a leading fellow of the Royal Irish Academy in Dublin, producing an extraordinary flow of scientific papers on diverse subjects. His paper, *Examination of the Supposed Igneous Origin of Stony Substances*, was one of a series criticising some of Hutton's weaker geological arguments (see Ch. 19), but also made the religious argument. Having previously dismissed Hutton's 'Theory of Rain', he began by praising him as having made an important meteorological discovery, but in his analysis of the *Theory of the Earth* he again lighted on Hutton's most memorable, but most mis-represented phrase. A process 'in *infinitum* was an abyss from which human reason recoils', he said.

> Into this gulph [sic] our author however boldly plunges; towards the end of his Essay he tells us, this earth is derived partly from one immediately anterior, and partly from another anterior to that again. In a word, to make use of his own expression, 'We find no vestige of a beginning.' Then this system of successive worlds must have been eternal; now succession without a beginning is generally allowed to involve a contradiction, therefore the system that forces us to adopt that conclusion must necessarily be false.[17]

This was a charge which had to be answered.

Chapter 16

. . . having found the most perfect evidence

NOT ALL OF the criticism came from far away. Some of Hutton's friends and Edinburgh associates had doubts about aspects of his theory. In 1788 John Walker, professor of natural history and fellow of the Royal Society of Edinburgh gave a public lecture entitled 'On the utility and progress of natural history and manner of philosophising'. After surveying the state of learning in botanical science he turned his attention to geology and to a favourite theme. After swipes at Descartes and Buffon, whose theories based on 'inadmissible postulate, would soon meet a sad fate', he made a more general point: theory had to be backed up with evidence.

> A probable opinion delivered merely as probable, or an ingenious conjecture to be decided by future experiment is pleasant to the philosopher and beneficial to philosophy, but this is quite different from attempting to arrive at certainty by a group of uncertain and wretched speculations.[1]

Walker's own reputation and career had been based on extensive fieldwork he had carried out in the Highlands and Hebrides. Although he did not mention Hutton by name, he certainly knew of his *Theory of the Earth* and may have been in the audience to hear it first announced. The challenge was clear: conjecture was not enough. If Hutton wanted to be taken seriously, he had to go into the field and gather evidence.

Walker was not the only doubter. Sir James Hall described being 'in almost daily warfare' with Hutton for three years.

I must own, that, on reading Dr Hutton's first geological publication, I was induced to reject his system entirely, and should
probably have continued still to do so, with the great majority
of the world, but for my habits of intimacy with the author; the
vivacity and perspicuity of whose conversation, formed a striking contrast to the obscurity of his writings.[2]

Despite the difference in their ages, Hutton, 59, had a soft spot for
Sir James Hall, 24, who was the son of his old university friend Sir
John Hall of Dunglass and his wife Maudie Pringle. Hall's mother
had died when he was two and his father when he was 15, but he
was taken under the wing of his uncle, Sir John Pringle (see Ch. 6),
who was president of the Royal Society of London. James was
educated in London, Cambridge and Edinburgh, where he had
studied under John Robison and Joseph Black. Hutton's farm at
Slighhouses was a short horse-ride from Dunglass, so he had probably met James as a child and again when he returned to Edinburgh
as a student. The young man had a lively interest in science and
geology. On a three-year Grand Tour of Europe, he had visited the
mines of Harz, climbed Vesuvius five times, explored Sicily with the
French volcanologist Déodat de Dolomieu, and had met Napoleon,
who was a student at the military academy. In politics and religion
he was said to be 'a declared democrat and avowed atheist'.[3]

Significantly, Hall had taken a firm stand on one of the major
scientific controversies of the age when he met Lavoisier in Paris
and was converted to his thinking. He brought his ideas back to
Edinburgh and in 1788 at three meetings of the Royal Society he
explained and argued for the 'new chemistry', based on Lavoisier's
ideas about the role of oxygen in combustion.[4] He was opposed by
Hutton who supported the older, widely accepted, phlogiston
theory, although he admired the work of Lavoisier and accepted the
presence of oxygen (which he called 'vital air') in burning. Hutton
persisted with phlogiston, which he defended in several subsequent
publications, because there were phenomena like heat and light
which could not be explained by the presence of oxygen alone.[5]

The appeal to phlogiston to explain aspects of the *Theory of the
Earth* was not the only objection. Hutton himself felt that he had
not properly answered his critics on the nature of granite, which

some naturalists believed was a 'primitive' rock dating from the Creation. Then there was his assertion that the consolidation of rocks from sand and other loose materials at the bottom of the oceans had been achieved by the action of compression under the weight of water acting with subterranean heat. Joseph Black had pointed out that if you heated limestone (calcium carbonate), it gave off 'fixed air' (carbon dioxide), but left a powdery oxide, rather than a consolidated rock. Playfair, although another close friend and disciple, had wondered about this apparent anomaly and Hutton was extremely vague in explaining where the heat came from.[6] There were also religious criticisms to be confronted and Walker's dismissal of his paper as just a theory, not based on experiment or evidence and therefore not properly scientific.

Over the next few years Hutton embarked on the most intense and productive period of his life, including several field trips to provide new evidence for his theory and a remarkable series of major new publications. These were designed not only to answer his critics on specific points (the nature of granite, or the criticisms made by de Luc), but to explore and explain his approach to science, philosophy and religion. He also embarked on some completely new topics: a 'Dissertation on Written Language as a Sign of Speech', which he read in two stages during 1786, and detailed explanations of his philosophy.

He had also begun to think about a supplementary paper on granite.[7] His argument was that far from being a primitive rock, granite was found in so many different colours and textures that it could not have been original. Some granites, called gneiss, were found in strata, and therefore had been formed by the laying down and consolidation of sediments on the seabed. Unstratified granite, such as porphyry, and basalt (called whinstone by Hutton), he conceded, originated from lava. In his mineral samples he had examples of granites forced into veins in layers of rock. He determined to find other examples in the landscape.

In September 1785 Hutton and his good friend John Clerk of Eldin set off from Edinburgh to visit the Duke of Atholl at Blair Castle. Clerk was one of Hutton's early confidents and supporters and an accomplished artist, who had also studied engraving and aquatint – valuable skills in producing printing plates.[8] Hutton intended not only to have a witness to his finds, but to be able to publish pictures of what they discovered. Clerk eventually produced nearly 70 drawings from expeditions he made with Hutton, although only a few were used in Hutton's subsequent book. The rest, plus some drawings by Sir James Hall and some by unknown artists, lay undiscovered among the papers of the Clerks of Penicuik until found by Sir John Clerk in 1968. Many of them were published in 1978.[9]

From his previous journeys around Scotland and the mental map he had built up of the geology of the country, Hutton knew approximately where he should look. He wanted to search along the fault line, where the granite mountains of alpine Scotland (the Highlands) met the sedimentary rocks that underlay the lowland areas. Clerk was also interested in geology and had some practical experience. His family had owned mines for many years and his father had written a dissertation on modern mining techniques. John Clerk himself had worked as a consultant, visiting and assessing pits on behalf of prospective buyers. In 1762 he had bought a half share in small loss-making coalfield at Pendreich, Lasswade, near Edinburgh.[10] Unable to afford to employ a manager, he did the job himself, descending the mine and learning about sedimentary rocks and their structure. Over time he made the pit profitable enough to enable him to buy a nearby estate at Eldin.

Travel in Scotland was still primitive and arduous, but Hutton was anything but reluctant. He was in late middle age, when many of his contemporaries had grown corpulent and comfortable, but he was still wiry and active. He was scientifically curious, but he also had a sense of adventure – the enthusiasm which shines out from the descriptions he later published was shared by his travelling companion. The main part of the trip to Blair Atholl was probably relatively easy, with reasonable roads to Perth and then following General Wade's military road, built in 1728–30 following the Jacobite Rising of 1715. The route took them via the town of Dunkeld, through the King's Pass, up the eastern side of Strathtay. It continued north through Ballinluig to Pitlochry before squeezing through the narrow Pass of Killiecrankie, scene of a historic battle in 1689. The road crossed the River Tilt at the Old Bridge to the north of Blair Atholl. The journey of 80 miles would have taken them two to three days.

The duke, who succeeded to the title when his father was drowned in the Tay in 1774, was a fellow of the Royal Society of London and 'a patron of science, particularly geology'.[11]

After leaving Blair Atholl the way would have become more difficult. They probably travelled on the backs of Highland ponies, rather than in coaches, and often the tracks were so narrow that they had to proceed on foot. Hutton and Clerk were determined to find what they were seeking, whatever distance they had to walk and no matter how steep the paths they had to climb. In the event, it was not difficult. The first discovery was made in Glen Tilt, close to the duke's hunting lodge, where they were 'entertained with the utmost hospitality and elegance'.

> His Grace now proposed to move us farther into the wilderness, and also to entertain us with the deer-hunting in his forest. We travelled up the Tilt, crossed the Tarf, which runs into the Tilt, and came to the other hunting seat of Fealar, the most removed, I believe, of any in Britain from the habitations of men. Here we were near the summit of the country, where the water runs into the three great rivers Tay, Spey, and Dee. The Duke was successful in killing three harts and one hind, all in excellent condition; and our curiosity was gratified in finding both the granite and alpine schistus in this summit of the Highlands, between Glen More and Glen Beg.[12]

Schist (schistus) was rock formed in layers from the consolidation of other rocks under pressure. What Hutton and Clerk found in the bottom of the riverbed, where the rock had been washed clean of earth and moss, were veins of lava granite, forced up into the strata which had been above.

> I here had every satisfaction that it was possible to desire, having found the most perfect evidence, that the granite had been made to break the alpine strata, and invade that country in a fluid state. This corresponded perfectly with the conclusion which I had drawn from the angular specimen of the Portsoy granite [in his collection].[13]

For this to have happened, the stratified rock must have existed before the granite was forced up into it. Granite was therefore the younger rock and could not have been a 'primal' rock and had not been formed at the Creation (although Hutton was circumspect enough not to spell that out). He was sure that his descriptions and Clerk's drawings proved his theory and any naturalists prepared to make the journey could now verify the evidence for themselves. Just to be sure he had a 400lb (180kg) boulder with an even more remarkable find – 'a vein which traversed both the mass of granite and broken schistus' – transported to his home and deposited in his garden.

The following year the two men set off again, this time heading south-west from Edinburgh to find the junction of the alpine granite with the schistus strata which Hutton knew might be found at the head of Loch Doon (called 'Loch Dune' by Hutton) in Ayrshire. They also wanted to search the coast of Galloway for places where the mountain granite met vertical schist strata. Though their journey was longer than it had been in Perthshire, the roads were better and they were able to travel part of the way by 'chaise', a light coach drawn by one or two horses. After some misinformation from the superintendent of one of Clerk's mines, they were again successful.

> We were extremely fortunate in finding what we looked for in two different places in Galloway; first, in the mountain of Cairn's muir, between two and three miles from the Ferry-town of Cree; and, secondly, in a little bay upon the seaside about mid-way between Covend and Saturness [now Southerness] point on the Solway firth. Here we were as much satisfied, as we had been the year before, that the granite had invaded the schistus or alpine strata, having not only broken and floated the schistus in every way possible, but in the last of those two places, we found the granite introduced, for some length, in small veins between the stratified bodies, giving every mark of the most fluid injection among the broken and distorted strata.[14]

Hutton was by now 60 and John Clerk 58, yet the two men had the excitement of schoolboys in discovering what they were searching for. Leaving their coach 'we ran with some impatience along the bottom of the sandy bay to the rocky shore which is washed by the sea, it being then low water'. And later: 'here we found the granite, not only involving the terminations of the broken and elevated strata, but also interjected among the strata, in descending among them like a mineral vein, and terminating in a thread where it could penetrate no farther. Mr Clerk's drawing, and a specimen which I took of the schistus thus penetrated, will convince the most sceptical with regard to this doctrine of the transfusion of granite.'

In August the following year Hutton set off again, this time for the isle of Arran, off the west coast of Scotland. He had hoped to visit the island the previous year, but was too late in the season to catch the mild weather. Clerk of Eldin was unable to go, so his son, also John Clerk, accompanied Hutton. Arran was especially interesting because it represented Scotland in miniature – the Highland fault line ran across the centre of the island, dividing the hills of the north from the flatter areas to the south. Hutton was again searching for examples of granite veins injected into the strata of older rocks. The two men searched unsuccessfully for several days, until one occasion venturing out alone, Hutton left his horse and climbed the side of a waterfall on Glen Shant Hill, on the foothills of Goat Fell, the large granite mountain which dominates the north of the island. There, where the water had washed away the moss and earth, he found what he was looking for and rode back to get Clerk to witness it and to draw it. This was confirmation of occurrences he had already found in Glen Tilt and in Galloway, but again he recorded it and arranged for a large boulder to be shipped back to Edinburgh.

But Hutton found something in Arran that he had not expected. For several years he had been searching for another sort of junction – this time not between sedimentary rocks (schist) and igneous rock (granite), but between strata of different ages. His theory suggested that stratified rocks, laid down at the bottom of the oceans, would periodically be expanded, pushed up and folded by subterranean heat, eroded down again and overlaid with new strata – all this happening over vast aeons of time. Finding a junction between the two sets of strata was for Hutton 'an object which I have long looked for, I may almost say in vain'.[15] He searched in the

Grampian mountains and in the southern uplands of Scotland and was beginning to think he would never find an example, but found one near Lochranza, in the north of Arran, 'where I had not thought almost of looking for it'.[16] On the shore, below high water mark, he saw two sets of strata, one above the other, both inclined at about 45 degrees, but in opposite directions so that the lower strata butted up against the upper strata.

> This, I think, is the only immediate junction that I have seen of the alpine schistus and what are commonly reckoned the strata of the Globe [sandstone and limestone] and acknowledge upon all hands to have been formed of matter deposited at the bottom of the sea.[17]

This was Hutton's first sighting of what became known as an 'unconformity'.* This was not new – more than a century earlier the Danish scientist Nicolas Steno had identified that all strata had originally been laid down horizontally at the bottom of oceans and that where they were angled as a result of some geological upheaval, the lower strata were the oldest. This was what Hutton was now looking at. An aspect of his genius was his ability to see small pieces of evidence – and he realised that at Lochranza he could only glimpse a fraction of what lay beneath – and imagine the process by which it had been formed. In his mind he began to envisage the geological history of Arran, its connection to the Ayrshire coast and islands and by extension how continents had been formed and destroyed. These were ideas he was to develop in the book he was planning, which would bring together the theories he had already published with the evidence he was now gathering.

After his return from Arran, Hutton visited a friend in Jedburgh, in the Scottish Borders. While walking along the valley of the River Jed, above the town, Hutton found an area where the cliff face had been worn away by the water. There he was surprised to see that although the top layers of strata were horizontal, below them the strata were vertical. He was so struck by this second 'unconformity'

---

* Hutton never used this term, which was coined by one of his critics, Robert Jameson, in 1805.

that he got John Clerk of Eldin to draw it and later used the illustration in his book. Again, his mind raced over speculations about how it had been formed and what it implied for his theory. Clearly the lower strata must be older and must have originally been horizontal, but heat and pressure had deformed them to the extent that they were now vertical. They had been eroded and again covered with newer horizontal strata.[18] Hutton later found other examples of vertical strata on the banks of the rivers Tweed and Teviot.

The following year, in late spring 1788, Hutton found what was to become his most famous 'unconformity', at Siccar Point on the Berwickshire coast. With John Playfair, he had accompanied Sir James Hall home from Edinburgh to his estate 35 miles (56 km) south-east at Dunglass. The area had a particular interest because Dunglass Burn, the stream which runs down to the coast, was the boundary between the counties of East Lothian and Berwickshire, but more importantly almost the boundary between the vertical and horizontal strata.

To the north-west of this burn and beautiful dean are situated the coal, limestone, marl and sand-stone strata; they are found stretching away along the shore in a very horizontal direction for some time, but become more and more inclined as they approach the schistus of which the hills of Lammermuir to the south are composed.[19]

The three men, Hutton, Playfair and Hall, examined the banks of several burns (streams), but the strata were not as clear as they were on the coast, where the sea had worn away cliffs up to 200 feet (60m) in height. They took a boat from Dunglass Burn and, with the weather fine and the sea calm, were able to sail close to the shore. With mounting excitement as they moved along the cliff they saw the horizontal red sandstone and marl strata of the lowlands lifting towards the schistus of the uplands:

But, at Siccar Point, we found a beautiful picture of this junction washed bare by the sea. The sand-stone strata are partly washed away, and partly remaining upon the ends of the vertical schistus; and, in many places, points of the schistus strata are seen standing up through among the sandstone, the greatest part of which is worn away.[20]

Playfair, Hutton's first biographer, was a first-hand witness to the event and his wonder at what he had seen and the way Hutton used it to tell the story of the Earth is evident in his account:

> On us who saw these phenomena for the first time, the impression made will not easily be forgotten. The palpable evidence presented to us, of one of the most extraordinary and important facts in the natural history of the earth, gave a reality and substance to those theoretical speculations, which, however probable, had never till now been directly authenticated by the testimony of the senses. We often said to ourselves, "What clearer evidence could we have had of the different formation of these rocks, and of the long interval which separated their formation, had we actually seen them emerging from the bosom of the deep?"
>
> We felt ourselves necessarily carried back to the time when the schistus on which we stood was yet at the bottom of the sea, and when the sandstone before us was only beginning to be deposited, in the shape of sand or mud, from the waters of a superincumbent ocean. An epoch still more remote presented itself, when even the most ancient of these rocks, instead of standing upright in vertical beds, lay in horizontal planes at the bottom of the sea, and was not yet disturbed by that immeasurable force which has burst asunder the solid pavement of the globe.
>
> Revolutions still more remote appeared in the distance of this extraordinary perspective. The mind seemed to grow giddy by looking so far into the abyss of time; and while we listened with earnestness and admiration to the philosopher who was now unfolding to us the order and series of these wonderful events, we became sensible how much farther reason may sometimes go than imagination can venture to follow.[21]

If they doubted Hutton's theory before, Playfair and Hall were now convinced.

Hutton's last field trip was to the Isle of Man in the summer of 1788. The Duke of Atholl, who had manorial rights over the island, asked Hutton and John Clerk to make a mineral survey. Though again they were entertained in fine style by the duke, Hutton found little to interest him. He wrote to James Watt: 'It is all schistus except a little bit of limestone on the shore at Castleton and a little sandstone at Peel. But I was much disappointed in not finding any body of granite, altho [sic] there are many blocks of it upon the surface, especially in one quarter.'[22]

In the same letter Hutton revealed that he was making a mineralogical map of Britain and asked Watt to find him, or draw, a geological map of the west country, the one part of the island he had not been able to visit. 'Could you help me to it? I mean what is called primitive schistus including granite. If you would just make a hand sketch of Cornwall, and draw a line including the primitive schistus. There is no occasion for accuracy, as we do not tell within an inch; it is enough to know that there is such a mass in Cornwall or Devonshire, and whereabout this lies or if there are two such masses there distinctly separate.' Watt did not draw a map, but gave Hutton a detailed description of the geology in his letter of reply.

Chapter 17

I was much weakened by it

THROUGH THE FIRST half of 1790 Adam Smith, who since his return to Edinburgh had been one of Hutton's closest friends, became increasingly weak. Henry Mackenzie, whom Hutton and Smith had taken to a meeting of the Royal Society of Edinburgh, wrote in June that the city had just lost its finest woman and in a few weeks would lose its greatest man. The finest woman was the beautiful Elizabeth Burnett, daughter of Lord Monboddo, whom Robert Burns had called 'the most heavenly of all God's works'. She died of tuberculosis aged 24. Smith, 67, was the greatest man: 'He is now past all hopes of recovery, with which about three weeks ago we had flattered ourselves,' wrote Mackenzie. The printer/naturalist William Smellie wrote: 'Poor Smith, we must soon lose him, and the moment in which he departs will give a heart-pang to thousands. Mr Smith's spirits are flat, and I am afraid the exertions he sometimes makes to please his friends do him no good. His intellect as well as his senses are clear and distinct. He wishes to be cheerful, but nature is omnipotent. His body is extremely emaciated, and his stomach cannot admit of sufficient nourishment; but, like a man, he is perfectly patient and resigned.'[1]

Smith had made Hutton and Black his literary executors and asked them to destroy his notes and papers, a task they avoided in the hope that his health might recover or he might change his mind. But a week before his death Smith became increasingly anxious and, too weak to do it himself, begged them to dispose of his files. Henry Mackenzie, who was present at the time, described the scene:

What a pity that he burnt these Vols, (of which I forget the number but it was considerable, I think) tho it seemed to be a measure he was fully resolved on & the performance of which gave him great satisfaction! I was at his Club, as the weekly meeting at his House used to be called, on the evening when this *auto da fe* was performed. I and one other friend remained in the outer room, while he with Drs Black & Hutton went into the inner apartment & burnt the Vols. When they returned his face, languid & pale as it was, beamed with Satisfaction. He felt the end nearly approaching & said with a complacent and good-humoured look 'I fear I must now adjourn this very pleasant Club to another World.'[2]

They burnt sixteen volumes of manuscripts without knowing what they contained. Smith had produced two major books, *The Theory of Moral Sentiments* and *The Wealth of Nations*, both of which had achieved wide circulation and influence and been republished many times. He had lectured on law and philosophy, given advice to policymakers and had been an assiduous public servant. But his only comment on hearing that his notes had gone was that he had done so little. 'I meant to have done more, and there are materials in my papers of which I could have made a great deal, but that is now out of the question.'[3]

In a brief appreciation of his friend for the Royal Society of Edinburgh, Hutton described Smith's reaction to the destruction of his papers: 'His mind was so much relieved that he was able to receive his friends in the evening with his usual complacency. They had been in use to sup with him every Sunday and that evening there was a pretty numerous meeting of them. Mr Smith not finding himself able to sit up with them as usual retired to bed before supper and as he went away took his leave of his friends by saying: "I believe we must adjourn this meeting to some other place."'[4] He died a few days later.

Smellie's prediction was wrong, Smith's passing was not widely mourned, even in Edinburgh. The diarist Henry Cockburn complained that the great and good of the city knew little about his work, beyond that he had been a commissioner of Customs and had written 'a sensible book'. It was the 'liberal young' who embraced his ideas, along with those of Hume, Robertson and the 'new chemistry' of Lavoisier.[5] The same was true of Hutton, whose most ardent supporters were Hall and Playfair. It is a reasonable assumption that outside the small group of his friends and the fellows of the Royal Society of Edinburgh, he was hardly known in the city and he was probably better known outside Scotland than within it. Among the people who ran the city – the merchants and politicians – if he was remembered at all, it was probably as his father's son. Playfair commented: 'He was little known, indeed, in general company.'[6]

With Smith's permission Hutton and Black had kept back some essays, which they published in 1795. They covered an extraordinary range – from a history of astronomy and five short articles on the external senses, to one on the affinity between music, dancing and poetry and another giving a comparison between English and Italian verses.[7] Some were obviously complete works, but others were fragments of unfinished longer documents. Smith appears to have exempted these from the flames because they gave an insight into his underlying philosophy. He believed that writers of 'didactical' discourse ought to 'deliver a system of science by laying down certain principles, known or proved, in the beginning, from whence we account for the several phenomena, connecting all together by the same chain'.[8]

Whether this influenced Hutton we do not know, but he began to prepare for publication several extended philosophical essays which, according to Playfair, he had been contemplating for some years. In quick succession he published *Dissertations on Different Subjects in Natural Philosophy* (1792), *A Dissertation upon the Philosophy of Light, Heat, and Fire* (1794), and in the same year *An Investigation of the Principles of Knowledge, and of the Progress of Reason, from Sense to Science and Philosophy*. These were among his longest and most detailed works and were put together at a time when he was ill with an affliction that could have killed him. Perhaps a sense of the possible approach of his end drove him to publish, so that he did not have to say, like Smith, 'I meant to have done more.'

Hutton was suffering severe abdominal pain which Joseph Black, acting as his doctor, diagnosed as acute suppression of the urine – probably caused by a stone in the bladder. Stones build up slowly, so this may have been troubling Hutton for several years and could have been the cause of him not being able to read the first part of his *Theory of the Earth* to the Royal Society of Edinburgh in 1785. The fact that he was able to undertake arduous field trips in the following three years suggests that either the condition was in remission, or that he bore the pain with immense fortitude.

By the autumn of 1791 the pain was so bad that Black decided to operate. The diarist Samuel Pepys had had a bladder stone removed in 1658, an operation performed without anaesthetic in the home of his cousin. He survived for almost 50 years, but Pepys was 25. Hutton was forty years older at the time of his operation 133 years later, but surgical techniques had not moved on by very much and measures to reduce the pain were still extremely primitive. In Hutton's case, if there was any attempt to dull the pain at all, it would have been by giving him brandy and warm water. Anaesthetic gases, such as nitrous oxide, were not in use until after Hutton's death.[9]

Black described the operation to James Watt, explaining that neither a catheter nor 'bougies' (surgical instruments for enlarging a tube) had been able to alleviate his suffering, and unlike Pepys, Hutton's stone was not removed. Perhaps Black thought that a more invasive procedure would have been too much for Hutton. Black had not practised surgery for some years and called on his cousin's son, the surgeon James Russell,[10] to assist him:

The bladder became much distended and very painful and the urine began to be absorbed and to affect his head when we made a puncture and introduced a flexible pipe above the *os pubis* [public bone]. This gave him some relief in the meantime – and a few weeks after the natural passage began to open again and is now restored nearly to its natural state, very little of the urine being now voided by the artificial passage. He was first taken ill in the beginning of August or not long after it and the danger he was in, together with the uncomfortable prospect that was before him gave me such a shock that I was much weakened by it.[11]

By the beginning of December, when Black wrote the letter, Hutton was sitting up, out of bed and working. To get over his shock at Hutton's state, Black had taken to riding, but his horse had fallen, crushing his leg and foot, which set him back. Hutton survived the procedure and, even if he was not in perfect health, was well enough to produce a substantial amount of new work in the following few years. A few months after his operation he wrote to Alexander Smellie, who was working with his father William in his publishing business, offering a paper for publication. With Smellie's encouragement he expanded it into a full-blown book.[12] He dedicated the *Dissertations on Different Subjects in Natural Philosophy* to Joseph Black, not as his surgeon, but as a scientist.

> The esteem which I have of your virtues and your knowledge and the long friendship which has subsisted between us afford sufficient motives for dedicating to you this volume. I have also another object in view. It is to make this public acknowledgment of that obligation which men of science owe you for your philosophical discoveries particularly for that of Latent heat – the Principle of Fluidity – a Law of Nature most important in the constitution of this World and a Physical Cause which like Gravitation, although clearly evinced by science is far above the common apprehension of mankind.[13]

The volume contained Hutton's published paper on rain, greatly expanded with examples and illustrations and answers to de Luc's criticisms. He also included an essay on wind. A second part outlined his defence of phlogiston in the debates at the Royal Society of Edinburgh with Sir James Hall. The third part was new – an exposition of a theory of matter – which included an essay on 'The Law of Matter and Motion, being an Enquiry into the Nature of Physical Body, its Constitution, Qualities and Accidents', and a second on 'The Law of Gravitation as a Principle of Natural Philosophy'. Playfair thought Hutton's theory of matter, examining the forces which attracted and repelled the smallest constituent particles within a body, was ingenious, but not wholly convincing.[14]

All these subjects – rain, wind, phlogiston, gravity – were germane to his *Theory of the Earth*, but he expanded and extended them, discussing, for example, the effect of gravity on the motion of the planets and the changing position of the axis of the Earth. A third dissertation looked at volume and how it was affected by heat and cold, and cohesion as a physical principle. Other essays examined the transmission of heat, the states of hardness and fluidity and the effects of electricity. At 700 pages, it was a substantial work. It contained no ground-breaking discoveries, but demonstrated that Hutton thought very differently about the nature of phenomena like heat and light than most of his contemporaries. He was also demonstrating that his *Theory of the Earth* had not come from a fanciful imagination, but was based on a thorough and detailed knowledge of the state of science.

A Dissertation upon the Philosophy of Light, Heat, and Fire, which followed two years later, began as a series of lectures to the Royal Society of Edinburgh delivered throughout 1794 and was then published as a book, although Hutton had to overcome his London publisher's lack of enthusiasm. He had the manuscript printed in Edinburgh (possibly by William Creech) and sent to Cadell and Davies, but by January 1795 he was complaining to them: 'I sent you a long time ago 200 copies of my quarto upon light and fire; I have heard nothing from you of their arriving; nor have I seen anything about their publication.'[15] Either the publisher had forgotten to inform the author that his work was in print, or they rushed it out after receiving his letter. The volume appeared with 1794 as the date on the title page, so Hutton was denied the opportunity of adding a 30-page supplement, which he had wanted to do.

The book was partly Hutton's response to two sets of experiments. The first, by Saussure on radiant heat, had been described in the second volume of his *Voyages dans les Alpes*. The second by Saussure's one-time student, Marc-Auguste Pictet, director of the Geneva Observatory, showed that 'radiation' from a flask of snow apparently could be focused by concave mirrors to cool another object some distance away. Although Hutton admired both men and their experimental approach, he disagreed with the conclusions they reached. The book grew from discussions of their findings to an eventual 326 pages, to include Hutton's ideas on the nature and effects of the elements, principally heat, which were important to his theory of the Earth.

In his discussion of the way different colour lights radiated different amounts of heat, Hutton was effectively describing infra-red light – that is radiation beyond the red end of the visible spectrum

– which he called 'obscure light'. The publication of his observations came five years before Sir William Herschel announced the same phenomenon to the Royal Society of London when he described 'calorific rays'. (The term infra-red was not coined until the late 19th century.) Hutton's method was more basic than Herschel, who had used a glass prism to split sunlight and had measured the rise in temperature with a thermometer. Hutton used a glowing coal as the source of red light and a flame as the source of white (or 'combined') light. It is not clear whether he used a thermometer to measure the heat given off by both sources, or whether he just relied on feeling the difference.[16] He also showed that 'invisible light' could be blocked by a sheet of glass, although visible light was not and that it could be reflected by a mirror.

William Greenfield, dean of the arts faculty at the university and minister of the High Kirk of St Giles in Edinburgh, read an early copy of the manuscript and wrote to Hutton that he had been 'staggered' by this finding, although he was sure that it would be confirmed by careful experiment.[17] But Hutton's pioneering work does not appear to have provoked much interest elsewhere. We do not know if Herschel read Hutton's book.

Hutton used heat to explain several of the processes which contributed to the cycles of his Earth machine. When describing how loose objects became hard strata, whether it was the bodies of countless billions of creatures which were turned into limestone, or sand and gravel washed down to the seabed which became sandstone, Hutton used subterranean heat as part of the explanation of how they had become consolidated into rock. But here he ran into an objection: it was well known that if you heated limestone you were left with a powder (quicklime). How was it then that Hutton's subterranean heat led not to powder deposits, but to rock strata? Black, in a paper published in 1756, had pointed out that heated limestone (calcium carbonate), gave off 'fixed air' (carbon dioxide),[18] but what if the gas could not escape? The resulting substance would be different. To bring about a situation where the gas remained captive, Hutton imagined the limestone trapped under the sediments above it and the weight of water in the oceans. So the consolidation was the product of heat and pressure acting together.

But where had the heat come from? Other than saying that the deep core of the Earth was a region which we knew little about, Hutton was vague about the source of the heat he relied on. That it existed was obvious to him, but he was not interested in its origin or its nature. What concerned him was its effect. His critics were by no means convinced that this source of heat existed at all, but even if it did, they pointed out that he expected it to behave in different ways at different times. Sometimes it was passive and consolidating, at others it was active, causing the stratified rocks above it to expand, buckle and occasionally to break. This was one of the major forces to which he attributed the creation of new continents from out of the ocean bed. To explain this difference Hutton resorted to another of Black's discoveries – latent heat and specific heat. Latent heat, he claimed, explained the fluidity of substances necessary before they could be compressed; specific heat explained the expansion and disruption of strata.[19]

This explanation made no sense to Hutton's critics. The conventional belief was that heat was a material substance called 'caloric' (see Glossary of scientific terms) and their frame of reference meant that they could not understand Hutton's arguments, which they saw as 'idle speculation without a theoretical basis'.[20] Also, since fire was a major source of 'caloric' and since volcanic eruptions spewed out fire, these critics thought of Hutton's theory as being necessarily based on the presence of internal fires within the Earth and questioned how such fires could exist.

Typical of the objections raised to the theory was a paper read to the Royal Medical Society in 1790 by John Thomson, who was a student at Edinburgh University. He sought to ridicule Hutton's *Theory of the Earth* for the demands it made on heat which had 'effects ascribed to it which it is evident that it could not produce'. If there were this source of heat within the Earth it would surely, over time, raise the entire planet to the same uniform temperature, he said. 'For such a fire not to have brought the whole Earth nearly to an equilibrium of heat, is contrary to every principle in philosophy, to everything we know of heat, and the laws of its communication from one body to another,' he asserted. For such a fire to exist it must have air and combustible fuel, but there was no air in the interior of the Earth, and the fuel, even if it existed at one time, would soon disappear. It was impossible to imagine that enough fuel existed anywhere within the Earth to keep a fire burning eternally, as called for in Thomson's view of Huttonian geology. Because of this, Thomson noted with sarcasm, whatever the vestiges of a beginning might be there was certainly some prospect of an end to Hutton's Earth.[21]

Thomson's view was not unique and similar criticisms had been made by de Luc and Kirwan, among others. They were explicable in the context of 'caloric', a mysterious fluid added to a body during heating. As more 'caloric' was added, the body expanded. When the body cooled 'caloric' was taken away and the body contracted. 'Caloric' always flowed from a hot body to a cooler one, but over the vastness of time imagined by Hutton, the central fires would constantly be adding 'caloric', so that the Earth would reach a uniform temperature and would always be expanding. Mountains would pop up continually. Given this mindset, Hutton's theory did indeed sound ridiculous.

Hutton thought of heat and light (and electricity) in a completely different way. To him they were not physical quantities, but manifestations of the energy flowing from the sun – he called it the 'solar substance'. They had neither weight nor volume, but were forces like gravity – although acting opposite: gravity was an attractive force; heat and light were repulsive forces. In his conception, when a body became heated the repulsive forces pushed it apart so that it expanded. When substances were burnt in a fire, they gave off heat and light, but it was easily possible to have heat without fire. Those arguing like Thomson and Kirwan that endlessly burning fires within the Earth were an impossibility because there was no oxygen and any combustible materials like deposits of sulphur, bitumen or coal would soon be used up, were irrelevant. Hutton's heat did not need fire, but he was not concerned to speculate on its nature or its origin. This lack of an explanation did not satisfy his critics.

Playfair and Hall both tried to provide answers later, pointing out that friction can cause heat, as can the sun's rays. Playfair also argued that the immense size of the Earth would be enough to conserve the heat within it, so negating the need for eternal fires, but these attempts to fill in the gaps in Hutton's argument, did not end the controversy, which continued after his death. Hutton also maintained that the intermittent effect of heat was the result of the interaction of the two opposing forces – the repulsive force of heat and the attractive force of gravity – sometimes one force would be in the ascendant, at other times the opposite force would predominate, although he was vague about how these interactions occurred. This, too, failed to silence his detractors.

Hutton was not a lone voice, his theories of heat and light had similarities with the ideas of Newton and Roger Joseph Boscovich, a diplomat and scientist from the Republic of Ragusa (modern day Dubrovnik), but Hutton's attempts to describe them and gain support were held back by his long-winded and repetitive writing style and his concentration on effects, rather than causes. They were so far in advance of the thinking of many of his contemporaries that attempts by Playfair to recast them at shorter length in simpler language largely failed to win them round.

Hutton's publications were reviewed by the *British Critic*, with its customary cynicism.

His two great works are filled with perpetual discussions on the abstract ideas of extension, magnitude, form, motion, inertia, hardness, expansion and the like on which subjects he attacks all received ideas without substituting one on which the mind can rest. Thus it is that he leaves those readers in total scepticism whose knowledge is but superficial or who depend on him for information.[22]

The sheer volume of words Hutton devoted to his explanations worked against him: 'The discussions generalisations and deductions of Dr Hutton on these three points [heat, light and combustion] are so long and diffuse that we could not easily pursue them without falling into the same prolixity.' Nevertheless, the magazine devoted a dozen pages to only the first section of his *Dissertations*.

A more sympathetic review of the *Dissertations on Different Subjects in Natural Philosophy* was carried by the *Monthly Review*, although it began with a similar admonition:

By labouring to attain great precision and perspicuity, he sometimes becomes tedious and obscure. The reader of these

papers and of his ingenious theory of the earth must be very
good tempered, if he bears patiently the dull formality of being
told over and over again what the author is going to prove, and
how he is going about it.[23]

The article, which apart from the criticism of Hutton's writing
style, was largely favourable, was written by Thomas Beddoes, who
taught chemistry at Oxford and was a friend of Erasmus Darwin,
Watt and Wedgwood. Beddoes was one of the leading proponents
of the kinetic theory of heat which rejected 'caloric', and so he
understood Hutton's reasoning.[24] His protégé Humphrey Davy
later proved that ice could be melted using heat generated by friction, which did not need fire.

Chapter 18

In some respects a foreign tongue

THE SHEER SCALE of Hutton's books, his ponderous writing style involving frequent repetition and attempts at explanation which often obscured rather than illuminated his meaning, concerned his supporters and irritated his detractors. It meant his work received less attention than it merited. This was true in his own time and still is today: his biographers have concentrated on *Theory of the Earth* and given scant attention to his other contributions to science or his general philosophy. According to the philosopher Peter Jones: 'The silence which greeted the publication in 1794 of the metaphysical treatise was, and has remained, almost complete.'[1]

The contrast between Hutton's apparently simple and concise verbal explanations and his ponderous written style may be because he was writing in what was to him almost another language. Unlike his contemporaries Adam Smith, David Hume and Joseph Black, he had not taken elocution lessons to make his broad Scots more easily understood in London. He was not alone. His friend John Clerk (and Clerk's son also John Clerk) and the senior judge Lord Braxfield clung to their native speech.

Alexander Carlyle, who had preceded Hutton at university, wrote: 'To every man bred in Scotland, the English language was in some respects a foreign tongue, the precise value and force of whose words and phrases he did not understand.' The Select Society, founded by Smith and Hume in 1754, engaged teachers 'qualified to instruct gentlemen in the knowledge of the English tongue, the manner of pronouncing it with purity and the art of public speaking'.[2] Hutton was not a member and probably would have refused lessons even had he been offered them. He did not expect a mass audience for his works, but he did want them to reach men of science and influence and probably felt he had to write in English.

His books are hard going, but they give us an insight into Hutton's views which we cannot get from any other source. The two books of *Dissertations*, published in 1792 and 1794, reveal much about his interest and knowledge of chemistry and physics as well as geology. They show that he kept up with developments in the natural sciences nationally and internationally. He was particularly interested in the experimental method and was dismissive of theories which were not based on sound evidence. He was a strong supporter of the discoveries of his friend Joseph Black, whom he worried would not get the credit he deserved – in particular he accused de Luc of trying to pass some of Black's discoveries off as his own.[3] He praised the work of younger scientists like Cavendish and Lavoisier, whose work, he said, 'is to be ranked among the greatest discoveries in physics'.

His defence of phlogiston, a theory now completely discredited, was not out of conservatism but because the 'new chemistry', while it explained much about combustion, left aspects which were still mysterious. For example, Cavendish's 1783 experiment in which he had produced water from the combustion of 'vital air' (oxygen) and 'inflammable air' (hydrogen) was accepted by Hutton, but he did not think it provided a full answer. Merely putting the two gases together would not lead them to combine. An ignition – a spark – was needed to set off the reaction, which generated heat and light as well as water. There was a missing element needed to explain the experiment fully. Until a better explanation was advanced, he stuck with phlogiston.[4] He rejected 'caloric' (called 'calorique' by Hutton) as a useful term by using similar reasoning. Caloric, like phlogiston, was another substance which could neither be measured nor observed except through its supposed effects. Where others needed to insert 'caloric' into a reaction, Hutton found more convincing explanations by using Black's concepts of latent and specific heat.[5]

His characterisation of heat and light as forces rather than physical substances was radical and enabled him to think about the expansion of materials when heated in an entirely new way. The addition of 'caloric' was not needed. Playfair obviously admired Hutton's capacity for original thought, but he was not uncritical, believing Hutton's descriptions of gravity and inertia, for example, were contrary to the laws of mechanics.[6]

Hutton's theory of matter (confusingly he used 'matter' to mean forces rather than particles), although superficially similar to that of Boscovich, marked him out as an original thinker. Playfair believed that Hutton had not read Boscovich and that his ideas owned more to the metaphysics of the German philosopher Gottfried Leibniz.[7] Others have noted similarities with the ideas of Immanuel Kant. Did Hutton know the work of Kant and Leibnitz? Playfair says he did not read speculative works, but 'absence of evidence is not evidence of absence', argues Peter Jones. Hutton was familiar with the works of Locke, Berkeley, Hume, Reid, Gregory, and Montesquieu, and 'had obviously read many French and British works on the origin and nature of language'.[8]

One of the frustrations in trying to get a full picture of how Hutton's ideas developed is that so few of his letters survive. On scientific matters we know he corresponded with Watt, Boulton, Darwin, Lind and Strange, but we do not know if he exchanged information and ideas with others in Britain and abroad. We have only a handful of letters from him to the Clerk family, the novelist Henry Mackenzie and the writer/publisher William Smellie. From a letter to Andrew Stuart, one of the protagonists in the Douglas Cause, a long-running dispute over inheritance which became a *cause célèbre*, we know he took an interest in current affairs.

Archibald Douglas, third Marquess of Angus and first Duke of Douglas, had no children and his sister Lady Jane Douglas was heir to his vast estate. In the event of her remaining childless, most of the duke's fortune along with a number of titles and ancient honours would pass to his kinsmen, the Dukes of Hamilton. In 1746 Lady Jane, then aged 48, secretly married Colonel John Stewart, a 60-year-old penniless soldier of fortune. The marriage seemed unlikely to produce children and after the wedding the couple fled to the Continent to escape their creditors. Two years later Lady Jane, now 50, announced that she was heavily pregnant. It was subsequently reported that on 10 July 1748 at the house of Madame Le Brun in Faubourg Saint-Germain in Paris, she had given birth to twin sons – Archibald and Sholto. Encouraged by the Hamiltons, the duke refused to recognise the boys as his sister's children and his heirs. He cut off her allowance and when the couple returned to Britain in 1751, Stewart was imprisoned for debt. Lady Jane and Sholto both died in 1753, and the young Archibald ended up in the care of his relative, Charles Douglas, third Duke of Queensberry.

In 1758 the Duke of Douglas, then aged 63, married Margaret Douglas, a middle-aged distant relative. She persuaded him to investigate the case for himself, which led him to reverse his previous decision and recognise Archibald Stewart as his heir. Ten days later, the duke died and Archibald, now aged 13, changed his surname to Douglas and inherited the duke's castles, properties and extensive lands in eight Scottish counties, with a fortune worth £12,000 per annum, the equivalent to £2.6 million in 2021.

Unsurprisingly, the Hamiltons did not accept this reverse. They sent the lawyer Andrew Stuart to France to investigate. He came back with the information that Archibald had in fact been born Jacques Louis Mignon, the son of a glass worker, and had been kidnapped in July 1748 by 'a lady, a gentleman and their maid'. He further claimed that the dead Sholto Stewart was the son of 'Sanry the Rope Dancer' and had vanished in similar circumstances. Stuart also reported that witnesses to Lady Jane's pregnancy could not be found, and that the couple had not stayed where they said they had.

In 1762, the Hamiltons launched an action in the Court of Session claiming that Archibald was an imposter. The case dragged on for years, with each side publishing long statements, incorporating letters, documents, witness reports, affidavits and citations of Scots and French law. A total of 24 lawyers read speeches to the 15 judges before whom the case was finally heard. The hearings lasted 21 days, making it the longest ever pleading before the Court of Session. The costs were estimated at £100,000, then an unheard-of amount for a legal action.

When the court gave its opinion, it was split down the middle, seven judges were in favour of Hamilton and seven for Douglas. The lord president, Robert Dundas, gave his casting vote in favour of Hamilton. Douglas' lawyers immediately launched an appeal to the House of Lords in London. The case, presided over by Lord Mansfield, the Scots peer who had become Lord Chief Justice of England, opened in January 1769 and lasted more than a month. A reputed £100,000-worth of bets were laid on the outcome. During the proceedings Andrew Stuart challenged one of Douglas' lawyers to a duel for calling him a liar. Pistols were fired, but both missed. The verdict, when it was finally delivered in February 1769, was unanimously in favour of Archibald Douglas.

The decision was wildly popular. In Edinburgh mobs smashed the windows of the lords of session who had opposed Douglas, and ransacked the Hamilton apartments in the Palace of Holyroodhouse. After two days of rioting, the military were called out to restore calm.

Following the case, Stuart published a series of open letters to Lord Mansfield, setting out his grounds for disagreeing with the verdict. Hutton's interest was probably piqued by the fact that the alleged births (or abductions) had taken place in Paris during the time of his studies there – although he would have been unaware of this at the time. He clearly followed the case closely and was surprised that the verdict appeared to go against the weight of evidence. He was one of many who wrote congratulating Stuart on his publication and in doing so stated his own principles. Although written in 1773, long before Hutton had published his own theories, it showed his realisation that the public statement of a contested thesis, however convincing, was likely to lead to attacks:

I read your letters to L[ord] M[ansfield] with great pleasure; I think they must be admired where they give no pain and will give much pain where they can give no pleasure. The publick [sic] justification of your character from attacks so unprovoked, so unmerited, is honorable, is compleat [sic]. Your defence of the sacred cause of truth violated in so conspicuous a manner by a person whose mind was neither weak nor uninformed, is as illustrious as the transgression was most infamous. Is there anything that truly may be called sacred amongst men but truth? Whatever in other respects are their opinions, however ridiculous, however uncertain, in this they all agree without hesitation that truth is to be held as sacred and falsehood an abomination, this principle is the very soul of society and is put in practice amongst the most worthy even where being neglected by the worthless.[9]

★ ★ ★

Hutton's books reveal him as a polymath. His system of the Earth could not be explained by mineralogy alone, it also needed a knowledge of meteorology, chemistry, mechanics and the physics of heat. Hutton would also claim it needed sound reasoning and a system of ethics and morality. These he tried to provide in his second book of 1794, *An Investigation of the Principles of Knowledge, and of the Progress of Reason, from Sense to Science and Philosophy.* In three volumes totalling over 2,000 pages, this was Hutton's longest work, its highly ambitious intention, announced in the first sentence of the Preface, was to 'investigate the nature or progress of human understanding'.[10]

He understood that the book's scale, its lofty purpose – and perhaps even his writing style – would not make it a bestseller: 'It is indeed only written for the few who make it their study to understand what they read and do not read merely to fill up a vacant hour in their oeconomy [sic]. But those few are all the world to an author; and, moreover these are the people who form the taste, morals and politics for all that is most valuable in this world, that is for the most civilised and the polished part.'[11]

In the first volume, running to 650 pages, Hutton examined in great detail his concepts of knowledge, ideas, passion, reason and action and attempted a definition of science ('the Conscious Principles which lead to Wisdom').[12] He set out his belief that knowledge is only acquired through experience, which was felt through the senses. In this volume, he did not specifically reject knowledge received from authority, particularly from biblical authority, but its absence was enough to make his position clear. Some philosophers have placed Hutton firmly in the tradition of British empiricism espoused by Locke (whom he had studied at university), Berkeley and Hume, although only the ideas of the first two were analysed by Hutton. The American philosopher J.E. O'Rourke also detected the influence of Aristotle's teleology – working backwards from effects, rather than forwards from causes – and an echo of the work of Kant, in Hutton's insistence that present processes can give us a guide to what happened in the past.[13]

That knowledge could only be based on observation, experiment, analysis and sound reasoning were familiar Hutton themes, but was the *Theory of the Earth* entirely derived from evidence? In his book *Time's Arrow*, the ecologist and science historian Stephen Jay Gould, argued that most of Hutton's fieldwork came *after* the initial presentation of his theory and was designed to validate his *a priori* assumptions.[14] On this reading, Hutton was not an empiricist at all.[15] Gould's conclusion, using selective quotations from Hutton himself, was controversial. Jean Jones, one of the most diligent of Hutton researchers, was provoked to scrawl on her own copy of Gould's book: 'this is a profoundly silly chapter. Gould is carried away by his own cleverness and pays little attention to fact.'[16] Jones points out elsewhere that Hutton was drawing on over 30 years of observation and recording of the geological landscape and from deductions made from close examination of the specimens in his mineral collection. This was undoubtedly true: until nearly the end of his life he was begging, borrowing and buying interesting and unusual mineral samples to analyse and record them.[17] His startling deductions about the injection of veins of igneous rock into existing sedimentary layers were made initially from mineral samples and only confirmed later by finding examples in the landscape.

But not everything in Hutton's theory could be deduced from his observations, nor tested experimentally. The consolidation of strata by pressure and heat, an important part of the cycle, sprang from Hutton's vivid imagination. He could not see it, feel it or measure it and he did not believe you could reproduce it artificially – it had to be taken on faith. James Watt and Sir James Hall both suggested that experiments could be designed to test this part of the theory. Watt wrote to Hutton in 1795:

> I have read the 1st vol. of your theory of the earth & without pretending to be an implicit believer, I see much to commend & admire. I do not believe even in Mechanicks [sic] without experiment, to which test I wish to bring all theoretical opinions if possible, & so I should have yours served. I think it possible to prove some parts of it by experiment though not all – I cannot pretend to find fires to melt marbles & flints, but I can to melt coals.[18]

Watt proposed heating lignum vitae, a hard, resinous wood imported from Central America, in the sealed barrel of a gun with a little water. The water would turn to steam, which would exert pressure, inhibiting the release of gas from the oil in the wood, 'which if it is not turned to coal will be something very like it'. The same apparatus could be used to test marble and chalk. Hutton, who was suffering pain and discomfort when he received the letter, did not respond and there is no indication that Watt carried out the experiment.

Hall also urged Hutton to test this part of his theory, but Hutton disagreed, arguing that it would not be possible to simulate the pressure or the intense heat needed so the experiment was bound to fail – and if it failed people would begin to cast doubt on this part of his hypothesis and, by extension, the whole theory. In the face of this objection Hall broke off experiments he had secretly begun in 1790 and only resumed them the year after Hutton's death. He restarted work in January 1798 and was undaunted by countless failures. In all he made 500 attempts to prove Hutton right, that carbonate of lime (calcium carbonate) would retain its hardness and not be reduced to powder if it was heated to extreme temperatures under pressure.

He tried different sorts of vessels of porcelain or glass before settling, like Watt, for gun barrels as the container for the limestone. Various methods were tried to seal the tubes – by welding, bolts or screws – with varying degrees of efficiency: on one occasion he was lucky to survive an explosion which destroyed not only the container, but also the furnace which was heating it. Hall managed to achieve temperatures of up to 27° Wedgwood* (approximately 2500°C) and had some partial successes, including producing a white marble which fooled the stonemasons he employed to polish it. Alas it fell into dust a few weeks later. In 1803 he began measuring the pressure applied to his samples, as well as the temperature.[19] Eventually, in 1811, 14 years after Hutton's death, he was able to announce to the Royal Society of Edinburgh that he had succeeded in producing hard stone, proving Hutton's speculation.

---

* The scale was an attempt by Josiah Wedgwood to measure temperatures above the boiling point of mercury (356°C). It was not very accurate.

This single result affords, I conceive, a strong presumption in favour of the solution which Dr Hutton has advanced of all the geological phenomena; for, the truth of the most doubtful principle which he has assumed, has thus been established by direct experiment.[20]

So the process of consolidation, conceived in Hutton's mind, was confirmed – at least to Hall's satisfaction – many years later by experiment. But to Hutton, his proposition, although he could not verify it when he published his theory, was no less valid than those that he could support by observed evidence. In volume two of his *Investigations* – even longer than volume one – Hutton elaborated his philosophy of science and in one section specifically defended the use of imagination in arriving at scientific truth. 'Real' and 'imaginary' were terms often contrasted and often used as synonyms for 'true' and 'false', he wrote. 'But imagination is considered as one of the sources of our knowledge and equally certain or believed with that of sensation.'[21] Certainty in establishing basic principles is possible in mathematics, but not in the physical sciences, he argued. Figures, magnitudes and proportions were innate ideas, but knowledge of physical things could only be acquired by experience through the senses.[22]

Hutton's immense discourse on reasoning from philosophical principles to reach scientific conclusions also considered equality and inequality, the difficulties in making judgements, the use of language, motion, space and time, common sense – and even madness. Much of the book outlined his general philosophy, for example in a section on the progress of reason Hutton examined doubt. Can a mind instinctively doubt something? he asked. His answer was no, a mind which was not conscious of knowing cannot doubt, because to doubt something required a judgement and judgement implied reason. Doubt, he asserted, is properly an operation of a scientific mind.

But contained within the vast mass of words are justifications for some of the key assumptions Hutton made in the *Theory of the Earth*. In a discussion of truth and probability, he defended his proposition that from our observation of the present we can infer what happened in the past and, by using the same reasoning, predict

what will happen in the future. This premise is absolutely fundamental to the *Theory of the Earth*, because we cannot observe
processes which happened untold ages in the past. We can only
infer them from what we see now and our belief that nature is
constant in its processes.

In the *Investigation* he gives an example: we know with absolute
certainty that the sun will rise tomorrow, so we can predict with the
same certainty that the sun will rise a thousand years from now. To
predict the same thing a million years from now might raise doubts
in the minds of some people, but it would be false to suggest that we
could not predict the sunrise with the same certainty merely because
the timescale has been lengthened. What we are actually saying is
that the same cause will have the same effect. Hutton refuted David
Hume in his reasoning on cause and effect[23] and argued further
that the converse is also true: we can deduce causes from knowing
their effects.

> So far as our reasoning concerning things, whether those that
> have been or those that are to be, shall be founded upon this,
> that the past and the future may be judged [of] from the
> present, conclusions may be formed from which there shall be
> no dubiety.[24]

In his section on evidence and principles, Hutton laid down the
three maxims which were the theoretical underpinning of his work:
that the laws of nature are established (and, although he does not
say it, the implication is that they do not change); that all things are
mutually related ('or else, how could we know them?'); and that
every effect has its proper cause. These, he confidently states, 'are
truly scientific principles'.[25] In the same section he answered the
critics who attacked him on religious grounds, calling him an atheist who denied the authority of the Bible – particularly the Creation
and Noah's Flood, described in the book of Genesis and attributed
to Moses. Hutton made clear his belief in a 'superintending being',
but his God was not the God of Moses, rather the logical outcome
of his own perception of nature as a perfectly constructed machine
(the emphasis in the following passage is mine):

All events being ordered according to certain rules, which are discovered in the observation of those actual things, herein is to be acknowledged *design*; in like manner, cause and effect being properly established by the rational mind of man, in the natural succession of things, here is to be acknowledged *wisdom*, which is the proper adapting of ends and means. Hence, as the universe in which those events are comprehended appeared to be without end or limit, so the *power* with which all this must be conceived as executed, will in reason be concluded as being in its nature infinite.[26]

Only through science could we discover the intention and wisdom of God:

That there is in every natural thing a clear revelation in which science must perceive a just establishment of order, a wise regulation of events and a benevolent cause of existence will, I hope, admit of no dispute; and without science no revelation could inform the animal man.[27]

If science was the only way to truly know the mind of God, was there any use for the Bible? Hutton allowed that there might be one:

The supposition of an oracular information, different from the voice of nature, must, to reasoning or philosophising men, appear absurd, although the belief of such a thing may be usefully employed in society for regulating the conduct of the vulgar or unlearned, who do not reason from principles of deep research or remote induction, who must be directed in their principles by the precepts of the wise and who may be happily influenced in their moral conduct by superstitious hope and fear.[28]

Although he did not name individuals it is probable that Hutton had Kirwan in mind when he wrote that it was not being able to distinguish between the superstition we take on trust as children and the belief in the existence of God arrived at through rational scientific principles which led men 'in their pious zeal to accuse philosophers of atheism'. But who had he in mind in the second part of that sentence? – 'and may also have led a philosopher to believe himself an atheist, when in reality he was only a rigorous inquirer into the validity of certain arguments'. Could it have been David Hume, whom Hutton accused of error in his reasoning elsewhere in this volume? There is no documentary proof of a friendship between Hutton and Hume, but they moved in the same circles and had a close mutual friend in Adam Smith. Hutton would have admired Hume's refusal to abandon his beliefs in the face of his attackers and the rigour of his reasoning, but would have been pained that from the same principles they apparently arrived at opposite conclusions on the existence of God.

Volume two includes many chapters discussing and defining God, the creator of a perfect system, the giver of all things good and the maker of pain and misery, which Hutton defends as necessary. There is a long section on 'nature', what we mean by the term and how we perceive it. Some of it echoes the metaphysics of Berkeley, whom Hutton had previously accused of sophistry and distrusting science.[29] Berkeley proposed 'immaterialism', that objects only exist in perception. Hutton asked:

> But where do these objects of our contemplation, objects considered as the works of nature, really exist? They exist in space and time; and where do space and time exist? In the mind of man, who knows these things. Without knowledge, to him they would be nothing; they are therefore in his knowledge and not in any other way.

But we should not conclude from this that they do not exist.

> It is in seeing this system of things intellectual, that we shall truly understand the system of material things, which are not false, although they only subsist within the mind of man, and there exist according to a pattern of which he has received from without, or of which he is informed by the author of his knowledge, that is by the hand of nature.[30]

Volume three, entitled 'Of Wisdom or Philosophy as the proper ends of Science and the means of Happiness', was presented as the conclusion of his arguments in the first two volumes. Although he praised Adam Smith's *Theory of Moral Sentiments* as 'a work that cannot be too much admired', he also took issue with Smith's assertion that sympathy of one human being for another is a fundamental principle. Hutton's style is also very different: whereas Smith, in lucid prose, argued a system of morality from observation of basic human interactions, Hutton's purpose and conclusions are more difficult to understand.[31] However, in the course of over 750 pages, he gives us his views on a range of subjects which otherwise we would know nothing about. They include broad areas such as politics and government, slavery, the role of women, globalisation, and injustice; and some more particular prejudices concerning intemperance ('pure folly') and drunkenness ('disgraceful and pernicious' but often pursued by men 'as if it were a blessing').

For the first time we see Hutton taking an interest in government. It was the role of philosophers, he claimed, to monitor the constitution of the state and to warn of any 'evil tendency' to deviate from sound policy. But in an age of revolution, Hutton was not taking a position alongside Thomas Paine on the side of the people in countering corruption in its leaders:

> The enlightened politician may either correct the fault, or oppose the growing evil with a proper remedy. But to suffer a popular cry, or the voice of the ignorant deluded by the sophistry of the interested and designing to guide the council of state respecting a political reformation, would be like the trusting of the wind, rather than the pilot, to bring the ship safe into harbour.[32]

He was agnostic about the form of government – it could be a monarchy, rule by the aristocracy or a republic – as long as it was wise in its legislation and virtuous in its administration. But once this wisdom and virtue were lost, the state could not endure.

> The first tyrant, who with cruel oppression shall put an end to anarchy, shall find himself established upon the throne of empire. He meets the wishes of a people preferring the arbitrary dominion of a peaceful sovereign to the hostile rage of a

conquering power, or wild disorder of discordant principles. But a system or tyranny once established in violence, and perfect virtue no more regulating government in justice, liberal men, aggrieved, only watch for an opportunity for retribution; and, with usurped authority they punish criminal despotism in the person of the tyrannical usurper.[33]

Two years before the book was published, the French monarchy had been overthrown by the revolution and Louis XVI executed. In the summer of 1794, when the book came out, Robespierre, who led the Jacobin *coup d'état* and instituted the Reign of Terror, was himself deposed and sent to the guillotine. Later in his book Hutton added that the words 'liberty' and 'equality' – two of the watchwords of the French revolutionaries – were not a guarantee of good government and often led to anarchy and confusion and thus came to mean 'violence' and 'injustice'.

He also considered corruption, deep-seated and widespread in British public life, with the use of patronage, pensions and sinecures. In a virtuous state, wrote Hutton, the wealthiest people engaged in the welfare of the nation because it was in their self-interest and because they desired the approval of their fellow men. But as states became larger and more complicated, the business of government became more specialist. Politics ('engaging in the great business of the nation') became a source of gaining wealth and with it the 'splendour of public offices' was increased.

> Here again begins the greatest corruption of our morals. For, when acquired wealth gives a title to pre-eminence, money will be sought merely for the sake of riches; riches will compensate for the want to virtue; and then nothing but philosophy can suspend the vilest degradation of mankind.[34]

He then touched on another hot political topic – slavery. There had been a reform movement since the 1760s – although it was 1807 before an Act was passed abolishing the trade transporting Africans to slavery on British West Indies plantations. In Scotland three cases had been brought by or on behalf of black slaves asking the Court of Session, Scotland's highest court, to rule their slavery illegal. Hutton may not have been aware of the first in 1756 because at that time he was farming in Berwickshire. The petitioner died before judgement could be given. A second case in 1769–70 ended with the death of the slave owner, with the slave gaining his freedom by default. The case provoked considerable public interest and Hutton may have taken an interest. The third case in 1778 was the first to result in a judgement. Joseph Knight, a slave brought from the West Indies to Scotland by his owner, John Wedderburn, petitioned the magistrates in Perth for his freedom, but was rejected. He appealed to the sheriff, who reversed the decision and declared that 'the state of slavery is not recognised by the laws of this kingdom, and is inconsistent with the principles thereof'. Wedderburn then appealed to the Court of Session, which upheld the sheriff's judgement and pronounced slavery 'unjust' and unsupported by Scots law.[35]

Hutton would have been well aware of another, local form of slavery, which bound mine workers to coal owners. In 1762 John Clerk had bought a half share in a mine at Pendreich, Lasswade. The pit had been sold by the Marquis of Lothian who proposed to move the workers, who were under indenture to him, to other mines he owned. In a celebrated court action several of the colliers took the issue to the Court of Session, claiming that their obligation was to the mine, not to the owner. But the judge ruled against them, finding, in the words of one pamphlet of the time, that the workers were 'but the personal Slaves of their Master, bound to work and serve him at any Coal he shall be pleased to employ them at'.[36] The slavery of colliers was not confined to the men alone. Often their wives and sometimes their children were also compelled to work in mines as 'bearers', carrying heavy loads of coal from the face to the base of the winding shaft, or in the case of drift mines, up steep inclines or stairs to the surface to unload their creels onto heaps. Some coal owners would bind children at birth, by paying 'arles', a small sum of money to the parents at baptism, with the child starting work at puberty. It was filthy, demoralising and back-breaking labour. The legal basis of the system was an Act of the Scottish Parliament of 1606, frequently amended to strengthen the rights of mine owners, by, for example, giving them the power to force 'vagabonds and sturdy beggers' into the pit.[37]

From the 1760s onwards the conditions of mine workers and their families became a scandal, although some mine owners were more motivated by the difficulty in recruiting labour for new pits than moral outrage. In 1772 Sir James Clerk, John Clerk's older brother, complained that the servitude of colliers was the real cause of 'the present great scarcity of hands we all justly complain of'.[38] However, more enlightened landowners did press for reform, resulting in the Colliers and Salters (Scotland) Act 1775. It noted that the Scottish coal workers existed in 'a state of slavery or bondage', ended the system for new pits and gave existing workers the opportunity to earn their way out of bondage, but it took another Act, after Hutton's death, to completely abolish this system in Scotland altogether.[39]

Adam Smith had condemned slavery in his *Wealth of Nations*,[40] but on economic, rather than moral grounds. Hutton, however, appeared to take a more principled stand. In a section entitled 'Philosophy of morals' he was unequivocal in his condemnation:

> Slavery and oppression are not part of government, or are not political government in any degree; no more than is the lashing of a generous horse in order to increase his speed. Certain political constitutions, indeed, may be more subject than others to be interrupted by such a disordering accident; but it is the proper business of government to prevent those evils in the state, by which the person and property of individuals are unjustly forced.[41]

The orders of the tyrant had to be obeyed because he had the power to enforce his will, but the will of the despot could be restrained by the virtue of the monarch and the 'natural humanity of man'. Those who enforced slavery were monsters, he added, and must be restrained by government, which must have in view the general good and not the gratification of any particular will.

Hutton considered various remedies to correct bad government and came up with a surprising solution: employ women in the service of the state. He stopped short of advocating full social equality for women, but since women had enjoyed no influence in the government of Britain since the death of Queen Anne, his was an unusual sentiment.

Nothing will contribute more to the perfection of a state than to have women employed as instruments in promoting public virtue and to see them valued for those accomplishments which best can make man happy; instead of being considered as only fitted for domestic service and for the idle entertainment of the little tyrant in the thoughtless moments of his life.[42]

To play their full part women needed to be informed and so should be educated in the art of government.

The office of women in the political state is of high importance; women form the manners of the children and they may form the morals of the youth, but for that purpose they should surely be *informed* [emphasis original]. They should be educated in useful knowledge and the purest principles, both of private and of public virtue they ought to understand the true forces of their rational, their independent happiness; and they should be made to see the wisdom and benevolence of that system in which man has his animal existence only for the purpose of his intellectual enjoyment – an enjoyment not temporary, not subject to corruption as are the sensual pleasures.

The education of women should not be reading Latin and Greek, or in studying abstract science, 'but in the art of government, which is the means of making mankind happy; and they should be accomplished in every species of learning that may lead to amiable and useful manners'. Later he added: 'It is in the wisdom of man to form women virtuous and intelligent; it is in the wisdom of women to form men honourable and temperate; and it is in the wisdom of philosophy to consider both female virtue and manly honour as bulwarks of the state which they adorn.'

This was a minority view in an age when, in Britain at least, universities, the learned societies, local councils, the House of Commons and the House of Lords were all closed to women. Schools for girls were rare: Edinburgh had the Merchant Maiden (1694) and Trades Maiden (1704) hospitals, both endowed by the money-lender Mrs Mary Hair*, but they only educated the daughters of 'decayed' or deceased merchants and tradesmen. Even Adam Smith, who had recommended the education of women in the *Wealth of Nations*, had not advocated giving women a role in government. Hutton's opinions seem closer to those of the feminist campaigner Mary Wollstonecraft, whose book *A Vindication of the Rights of Woman with Strictures on Political and Moral Subjects* had appeared in 1792.

Playfair tells us very little about Hutton's women friends, other than that he enjoyed the company of 'several excellent and accomplished individuals of both sexes'. We know he got on well with the wives of his friends James Watt and Matthew Boulton and in their letters to him or to Joseph Black they asked after him. He began his letter to George Clerk with a message to Clerk's wife, Dorothea Clerk Maxwell, regretting that he had not written to her personally and making a promise to write with 'a wonder deal to tell you'.[43] He lived with his sisters, Isabella and Jean (Sarah had died in 1790 and Jean was to die in 1795), but other than that they were 'excellent women who managed his domestic affairs', we know little of them or their views. And, of course, there was the mother of his child, about whom we know nothing. Like so much of Hutton's life, the women who may have helped to shape his opinions remain a mystery.

---

* Now better known by her maiden name, Mary Erskine.

Chapter 19

The doctor will not easily submit

THE PRESS NOTICES of Hutton's *Investigations* were predictably critical. The *English Review* could hardly be bothered to read it, even though it was 'the boldest and most singular inquiry that this age has produced into the philosophy both of nature and of mind':

> The time that we can afford to its perusal is altogether insufficient for any formal examination of its truth and the limits within which we are confined prevent us even from stating the whole of those doubts or objections which have occurred to us to many of the opinions and reasonings of the author. Of works of this magnitude it is posterity alone that is the judge.[1]

There then followed a dozen pages, largely made up of direct quotations. Tobias Smollett, writing in the *Critical Review*, was blunter in his criticism of the extent and style, although his appraisal was more thoughtful:

> Five times has the printer reminded us of the length of time this work has been upon our hands, five times have we assailed the three quarto volumes but in vain. We have been repulsed at each attack and at this moment we feel ourselves unable to do justice to ourselves, to the author, or to the readers. To what can this be owing? . . . The writer thinks prolixity, of the evils he might incur in his works, the least and we are dragged without mercy through the science of philosophy and the philosophy of science, through the theory of idea and the idea of theory, till we know not whether we stand upon our head or our heels.[2]

Only the *Analytical Review* gave Hutton a lengthy, detailed and largely more positive assessment, spread over two issues and ending with the conclusion:

> This investigation alone would justly intitle [sic] him to a distinguished rank among the philosophers of the present age and we dismiss it with returning him our grateful acknowledgements for the pleasure and the improvement we have reaped from its perusal.

However, the praise was tempered: 'The form in which it is presented we are sorry to say is in many instances not only inelegant but to every reader of taste highly offensive and forbidding . . . We mention however these circumstances with no intention to detract from the general merits of the work which we deem deserving of the highest commendation, or to derogate from the author's well-founded claims to literary honour.'[3]

It is unlikely that Hutton was swayed by either the criticisms or the praise. Again, his ponderous writing style had inhibited intelligent discussion of his theories. We can only speculate why he did not seek help from a friendly editor like Black or Playfair to help him achieve a more accessible style. Perhaps he realised that his time was running out. Rather than honing and polishing books he had already written, he had a much more urgent task in hand – the publication of a new and greatly expanded edition of his *Theory of the Earth*. The operation performed by Joseph Black had given him a few years of relative good health; in January 1794 Black was able to tell James Watt that Hutton was 'in excellent spirits',[4] but by June his old trouble was back. Black again had to operate and reported that Hutton had been confined to bed – 'free from pain and in a condition to amuse himself with his books and writings and much better for the attention of his friends' – but there was a gloomy prognosis: 'there is as yet no appearance of his recovering so well as he did before'.[5]

Hutton's friends had been urging him to publish a more complete version of his theory, including the evidence he had collected on his field trips and the illustrations by John Clerk, which had not so far been published, but he had been putting off the work and they feared it might not appear in his lifetime. The spur he needed was reading a report of the attack made on him by Richard Kirwan (see Ch. 15). The very next day, Playfair tells us, Hutton began to prepare the proofs for the printer. This was despite his failing health. Black wrote to Watt:

> Dr Hutton has been close confined to the house for this twelve-months and is now without hope of relief from his confine-ment, but he enjoys good spirits and is constantly employed in writing and publishing. Mr Kirwan has attacked his *Theory of the Earth* and the doctor will not easily submit to such an antagonist.[6]

Hutton was being cared for by his sisters, but in August 1795 Jean died at the age of 70. This left only Isabella, who was 72, but apparently in robust health. Black was regularly making the ten-minute walk from his home in Nicolson Street to St John's Hill to see Hutton. His own health had not been good, but he had shaken off a persistent respiratory complaint which saw him coughing up blood by following a strict regime of daily 'currying' (vigorously towelling his body with a coarse cloth) and an almost vegetarian diet. He had recovered from the riding accident when his horse fell and crushed his leg and foot, or at least he no longer mentioned it in his letters to Watt. Hutton and Black had been editing Adam Smith's essays, which were finally published in London by Andrew Strahan and Thomas Cadell. The contract, negotiated by Henry Mackenzie, involved a fee of £300, but Hutton and Black received nothing, the money going to Smith's nephew and heir, David Douglas.

In September, Black reported to Watt that his patient enjoyed 'a wonderful degree of courage and good spirits'. Hutton had sent the first two of four planned volumes of the *Theory of the Earth with Proofs and Illustrations* to the printer. He wrote to Cadell (now in partnership with William Davies) on 28 September to tell him that 1,187 pages of text had now been printed, but pages with the six illustrations (four quarto and two folio)* were still to be produced. Some 200 copies were to be bound and sold in London at 14 shillings (70p), with a further run being sold in Edinburgh by his local

---

* Quarto was approximately 24cm x 30cm and folio 30cm x 48cm.

publisher, William Creech. This was not the complete work: Hutton
wrote to Cadell again on 6 October to tell him that another volume
was to follow 'with many plates'.[7] By December, Watt was able to
tell Hutton that he had read his dissertation on heat, light and fire
and was beginning the *Theory of the Earth*.[8] Hutton may have sent
him the printed pages and Watt had them bound into book form
himself because it does not appear that the work was on general sale
until early 1796.[9] By this time Hutton was enduring frequent fits of
pain from the stone in his bladder, but was 'very lively at times',
according to Black.

The first 200 pages of the new book were a restatement of
Hutton's theory as it had been published in 1788, although there
were a number of minor revisions.[10] However, in the following
pages Hutton laid into his critics, particularly Kirwan. A theory
which was founded on a new principle and overturned the received
opinions was bound to meet opposition from 'the prejudices of the
learned and from the superstition of those who judge not for themselves in forming their notions, but look to men of science for their
authority'.[11] Hutton wrote that he accepted criticism and welcomed
the opportunity to defend his claims for it was when truth and error
were forced to struggle together that progress was made in science,
but Kirwan he accused of a 'train of inconsequential reasoning'
leading to a 'crude and inconsiderate notion'.

> The observations which he has made appear to me to be
> founded on nothing more than common prejudice, and
> misconceived notions of the subject. I am therefore to point
> out that erroneous train of reasoning, into which a hasty super-
> ficial view of things, perhaps, has led the patron of an opposite
> opinion to see my theory in an unfavourable light.[12]

Hutton went on to accuse Kirwan of misrepresenting his arguments, of inconsistency and of misreading the research of others.
He rebutted the criticisms of the erosion of rocks, the creation and
the destruction of soils, the formation of new strata under the sea
and the mineral structure of Scotland. Kirwan had claimed it was
largely granite; Hutton countered that 'there is scarcely one five
hundred[th] part of Britain that has granite for its basis'.

I had examined Scotland from the one end to the other before
I saw one stone of granite in its native place, I have moreover
examined almost all England and Wales, (excepting Devonshire
and Cornwall) without seeing more of granite than one spot,
not many hundred yards of extent.[13]

The importance of granite was that Kirwan considered it a 'primary'
rock – that is one having existed since the Creation, which he
believed had followed the pattern described in Genesis. Hutton had
shown that not only was granite rare, but that it was a younger rock
than the sedimentary layers which often covered it. This he had
demonstrated by finding examples, such as in Glen Tilt, where
granite had forced its way into cracks in the older sediments. The
argument between the two men came down to Hutton's belief that
the Earth had evolved slowly over extremely long periods of time
(and was still evolving) and Kirwan's adherence to the 'Mosaic'
narrative – the biblical story attributed to Moses.

The difference between the two views was summed up in Kirwan's
attack on Hutton's phrase 'we find no vestige of a beginning'. Like
many other critics, he interpreted this as an assertion that the world
was eternal – and therefore had not been created by God. Essentially
Kirwan was accusing Hutton of atheism. Hutton countered:

Our author might have added that I have also said '*we see no
prospect of an end;*' but what has all this to do with the idea of
eternity? Are we, with our ideas of *time* (or succession), to
measure that of eternity, which never succeeded anything and
which will never be succeeded? Are we thus to measure eternity, that boundless thought, with those physical notions of
ours which necessarily limit both space and time? and, because
we see not the beginning of created things, are we to conclude
that those things which we see have always been, or been without a cause? Our author would thus, inadvertently indeed, lead
himself into that gulf of irreligion and absurdity into which, he
alleges, *I have boldly plunged.*[14] [Emphasis original]

Hutton then set out his deist view: that God was behind the complex
but elegant system the *Theory* described. He turned back onto

Kirwan the accusation that he was underestimating the 'wise designer' of the Earth.

> Would he deny that there is to be perceived wisdom in the system of this world, or that a philosopher, who looks into the operations of nature, may not plainly read the power and wisdom of the Creator, without recoiling, as he says, from the abyss? The abyss, from which a man of science should recoil, is that of ignorance and error.

But not all Kirwan's criticisms were easily dismissed. Hutton claimed subterranean heat as one of the agents helping to consolidate strata, but what was the source of this heat? If it came from fire, what fuelled the fire and if it burned for all eternity, why did the fuel not run out? In his *Theory* Hutton had not attempted an explanation of what provided the heat. He knew he was on weak ground. It was for others to explain the fire, he maintained; it was enough that he had proved that heat was present: 'For though I should here confess my ignorance with regard to the means of procuring fire, the evidence of the melting operation, or former fluidity of those mineral bodies, would not be thereby in the least diminished.'[15]

Hutton spent more than 60 pages of his new book refuting Kirwan, before turning to other theorists, including Buffon and de Luc and concluding that they were wrong. He was at pains to refute two widely held theories: that there existed 'primal' rocks and that some great deluge (like Noah's Flood) had once engulfed the Earth and was responsible for many geological features. A rather amusing claim by the Prussian naturalist Peter Simon Pallas that the remains of mammoths and rhinos which he had discovered in Siberia could be explained by floodwaters bringing the bones from Africa. Hutton concluded his geological argument with the statement: 'at the same time we would ascertain this fact, that formerly the Elephant and Rhinoceros had lived in Siberia'.[16] The book also gave details of his field trips and the findings he had made which supported his theory. He ended his first volume with three chapters on coal, which drew on his own observations from across Scotland and England and on his first pamphlet on the difference between culm and coal.

In his second volume, Hutton applied his theory to the evolution of continents. He used the published observations of Saussure and de Luc, against their own theories (which favoured the biblical narrative) and in support of his own. He considered coastal erosion and stated that England had once been joined to France (echoing Desmarest's student essay), Ireland joined to Britain, Orkney with Scotland, Shetland with Norway. He quoted Thomas Jefferson on the natural stone bridge of Virginia as an illustration of the power of water to change the landscape. He described the way glaciers can move huge boulders, and gave examples of other geographical phenomena from Sumatra, Peru, Spain, Colombia, France, and Italy. These he must have gathered from his wide reading or directly by corresponding with travellers. But, he asserted, the power of water to dramatically alter the landscape took a very long time, it was not the result of one catastrophic flood. Near the end of this volume he neatly linked his theory of the circulation of rock – from its formation as sediment, to its upheaval to form mountains and land, to their erosion by weather, their transport in rivers to the sea, to be again deposited as sediment – to his medical dissertation, written over 40 years before.

> We live in a world where order everywhere prevails and where final causes are as well known, at least, as those which are efficient... Thus, the circulation of the blood is the efficient cause of life, but life is the final cause, not only for the circulation of the blood but for the revolution of the globe.[17]

And, as a final condemnation of those who believed in a single catastrophe, or some sort of supernatural power, rather than a gradual evolution, he wrote:

> Not only are no powers to be employed that are not natural to the globe, no action to be admitted of, except those which we know the principle, and no extraordinary events to be alleged in order to explain a common appearance, the powers of nature are not to be employed in order to destroy the very object of those powers; we are not to make nature act in violation of that order which we actually observe and in the subversion of that end which is to be perceived in the creation of things.[18]

Hutton's arguments did not silence his critics. In an unsigned review which ran over three editions of the *British Critic* and occupied 48 pages, de Luc summarised the theory in seven points and then attempted to rebut it point by point.[19] Referring to Hutton throughout as 'Dr H', he alleged that Hutton had taken the facts observed by other authors and misrepresented them to support his own theory. Where Hutton had quoted Saussure and Kirwan, de Luc took their side. He concluded with what he believed was a damning condemnation:

> We are now arrived at the end of a laborious task, which nothing but its importance could have determined us to undertake . . . it has given us the opportunity of stating some fundamental points in Geology; a science of which it is become necessary to have a just idea, in order to avoid falling into the paths of those, who, fancying they have discovered the secrets of nature, without having studied it with the attention requisite for such a subject, would make us forget that sacred history, which, at the same time that it gives us the first true information on the origin of the universe and the history of the earth, teaches us the purpose of these Revelations from the author of nature; that of prescribing to men precise duties, and giving a certain, but conditional, foundation to their future hope.[20]

By the time Kirwan returned to the attack in November 1796 the political atmosphere in Britain had become even more heated – in France Louis XVI had been executed and during the Reign of Terror Robespierre had passed a series of anti-religious laws, before himself being sent to the guillotine. There were constant rumours of imminent invasion of Britain by a revolutionary army. Kirwan's new paper again countered some of Hutton's geological arguments, but explicitly denied Hutton's claim that science had an exclusive right to consider the means by which the Earth was created and directly accused him of atheism.

> Recent experience has shewn that the obscurity in which the philosophical knowledge of this state has hitherto been involved, has proved too favourable to the structure of various

systems of atheism or infidelity, as these have been in their turn to turbulence and immorality, not to endeavour to dispel it by all the lights which modern geological researches have struck out. Thus it will be found that geology naturally ripens, or (to use a mineralogical expression) *graduates* into religion, as this does into morality.[21]

A third assault came in another defence of Genesis by Philip Howard in his book *The Scriptural History of the Earth and of Mankind*, published in 1797. The book had previously been published in French as a series of letters in which the author dismissed the theories of Buffon, but following the appearance of Hutton's new statement of his theory, Howard published an English translation, with added letters taking the side of Hutton's critics, notably de Luc. Howard stuck rigidly to the biblical narrative, including accepting the age of Methuselah at 900 years, which he ascribed to a golden age of abundance following the Deluge. The substance of his book could be summed up in his phrase: 'Profane-written history is indeed modern and fallacious.'[22] He did, however, commend Hutton's discoveries of the different properties of coloured light. In its review, the *British Critic* commented:

In this refutation, which we cannot allow ourselves space to insert, the author most clearly shows, that all they who, in their endeavours to trace the history of the earth, have departed from the text of Scripture, have, at the same time, departed from the facts, and from the demonstrated principles of physics. Particularly that it is a vain attempt to endeavour to pass by the Deluge, such as it is described by Moses, and at the very period where it is placed by him, in any explanation of the present state of the globe.[23]

The Creationists were not to be defeated either easily or quickly.

Chapter 20

Death only as a passage

HUTTON WOULD HAVE been delighted to receive bound copies of his final theory. He was now housebound, suffering periods of severe pain and looked after in his home in St John's Hill by his sister Isabella, his servants and his doctor, James Russell. His devoted friend Joseph Black, who was now living a short walk away in College Wynd, called regularly (and sent health reports to their mutual friend James Watt), but was only two years younger, not in perfect health himself and not maintaining a professional medical practice. The gregarious Hutton could no longer walk around Edinburgh to drop in on his acquaintances, but he would not have lacked for visitors. They almost certainly included Andrew Balfour, the son of his dead cousin, to whom Hutton had acted as a second father. The young man was now following in his mentor's footsteps, having progressed from the High School to study medicine at the university. Hutton presented the young man with one of the first copies of his book.*

Hutton's two disciples, Professor John Playfair and Sir James Hall, would keep him abreast of scientific news and his business partner, John Davie, when visiting the city from his estate in West Lothian, would have kept him up to date with the commercial gossip of Edinburgh and agricultural developments. Their partnership was apparently still providing Hutton with an income. John Clerk, his companion on his recent field trips, probably called to discuss the illustrations he was to provide for the final parts of his friend's theory.

---

* It is now in the library of the Royal Botanic Garden Edinburgh, presented by Balfour's son John Hutton Balfour, who became Regius Keeper of the garden in 1845.

With his medical knowledge, Hutton probably knew he was on his death bed with little chance of leaving it alive, but the house would not have been in premature mourning. Hutton did not fear death and did not see it as a final end. His view, expounded in his philosophical treatise, was encapsulated by Playfair:

> Death is not regarded here as the dissolution of a connexion between mind, and that system of material organs, by means of which it communicated with the external world, but merely as an effect of the mind's ceasing to perceive a particular order or class of things; it is therefore only the termination of a certain mode of thought; and the extinction, not of any mental power, but of a train of conceptions, which, in consequence of external impulse, had existed in the mind. Thus, as nothing essential to intellectual power perishes, we are to consider death only as a passage from one condition of thought to another; and hence this system appeared, to the author of it, to afford a stronger argument than any other, for the existence of the mind after death.[1]

This belief, and Hutton's natural ebullience in the company of stimulating intellectual argument, would have made for a lighter atmosphere than might be expected. And when there were no friends to provoke and amuse, he was engaged in his next favourite occupation – working.

Having seen a fuller explanation of his grand theory launched in public, Hutton still echoed the dying words of his friend Adam Smith – he felt he could have done more. Despite his failing health he maintained a punishing work schedule. The published theory was still not complete and, as he had warned his publisher, a third volume was to follow 'with many plates' (illustrations) and after that a fourth. He gave a preview of what these additions would contain in the final paragraph of volume two, when he promised they would provide more geological facts and cover philosophical ground:

> In this physical dissertation, we are limited to consider the manner in which things present have been made to come to

pass, and not to inquire concerning the moral end for which those things may have been calculated. Therefore, in pursuing this object, I am next to examine facts, with regard to the mineralogical part of the theory, from which, perhaps, light may be thrown upon the subject; and to endeavour to answer objections, or solve difficulties, which may naturally occur from the consideration of particular appearances.[2]

By 1796 Hutton had largely finished nine chapters of volume 3, dictating to an amanuensis, possibly his sister Isabella. He then had the individual chapters bound in boards to make editing easier, went through them and made notes and changes in his own hand-writing. However, this was not the philosophical essay promised at the end of volume 2, but longer descriptions of his field trips to Glen Tilt, Arran, and the south of Scotland, and particularly of his search for granite. He left gaps where it was intended that etchings made from John Clerk's drawings would appear. There was also another refutation of the theories of Saussure, and new evidence from the Pyrenees and Calabria which he claimed supported his theory.[3] The volume finished with a detailed geological history of Arran. He did not send the new work to the printer, but put the manuscript aside, perhaps waiting until he had written the final volume, or waiting for Clerk to produce the illustrations.

Hutton also had another project in course of preparation – a trea-tise on agriculture. It had been in his mind since his farming days in Berwickshire – he called it 'the study of my life'. He had started work on it in 1794 and added to it in the next two years. By his last year he was working on a fourth section and annotating the work he had already done.[4] In a handwritten preface he declared his aim was not to write a popular book, but to bring together practical husbandry with science – 'to give the principles with the practice, and the practice with the principles'.[5] Most of the rest of the book was dictated, although there are corrections in Hutton's handwrit-ing in the first of the two volumes and the first few pages of the second. They begin in ink and then continue in pencil, although some are feeble perhaps indicating his failing strength.[6] Most of the contents bring together his observations of farming methods in East Anglia and in the Low Countries of Europe, together with his practical experience in Berwickshire (see Ch. 6). But he also included lengthy sections on plant and animal breeding and clearly explained the principle of natural selection as part of the evolutionary process. He first framed the issue in general terms:

If an organised body is not in the situation and circumstances best adapted to its sustenance and propagation, then, in conceiving an indefinite variety among the individuals of that species, we must be assured, that, on the one hand, those which depart most from the best adapted constitution, will be most liable to perish, while, on the other hand, those organised bodies, which most approach to the best constitution for the present circumstances, will be best adapted to continue, in preserving themselves and multiplying.[7]

Then he gave two examples. If speed and keen sight were the necessary attributes for hunting dogs, then those individuals with those abilities would thrive, sustain themselves and continue the race. Slower dogs and those with deficient eyesight would starve. Similarly, if a sense of smell was the crucial attribute, then evolution would 'produce a race of well-scented hounds, instead of those who catch their prey by swiftness'. He also gave an example from plants: 'the most prosperous plant must be that which will furnish, with its maturated seed, a vigorous race of fertile plants; and, these will be more and more accommodated, in the varying power of vegetation, to the soil and circumstances in which they grow.'

Man had perceived these laws of nature and harnessed them to his own end in selective breeding of animals and plants.

The ingenuity and art of man is boundless; nothing can happen in nature that he may not notice; and nothing can be the subject of his observation that may not, in his science, be employed for some valuable purpose. Who would have thought, that his study of a fly should have contributed to the cultivation of his fruit? By the ingenuity of man, the destructive influence of pernicious insects may surely be prevented upon many occasions; but, to employ an insect for the spinning of his thread, and for the ameliorating of his fruit, even when he is ignorant

of how that insect operates for his purpose, is an example of
the acuteness of the human genius, in observing nature, and
reasoning from effect to cause, which perhaps has not its equal.

Hutton noted that the farmer-scientist Robert Bakewell bred sheep
for the quantity and quality of their meat, whereas in Lincolnshire
they bred sheep to maximise the value of the fleece. Hutton believed
that it would be possible to produce a breed which gave both wool
and meat. He also gave examples of efficient plant breeding to
maximise the benefit from crops. Although his *Elements of Agriculture*
was never published in his lifetime, his explanation of natural selection predated Charles Darwin's *On the Origin of Species* by more
than 60 years. It is unlikely that Darwin had read Hutton, but we
know that he was influenced by the work of his grandfather,
Erasmus Darwin, who was a friend and correspondent of Hutton
and who had expressed his ideas about genetic change in his prose
and poetry. Whether Hutton and the older Darwin ever discussed
their theories of natural selection we can only guess. Charles Darwin
also read Charles Lyell's *Principles of Geology*, which was strongly
influenced by Hutton's ideas.

Hutton's manuscript was global in its scope, discussing, among
other topics, the influence of polar conditions on the weather of
northern temperate countries. He drew explicitly on the work of his
friend Joseph Black and the concept of latent heat to explain the
effect of melting ice on climate and drew on his own experience
and observation to illustrate the way in which the sun warms the
soil and the effect this has on plant growth. He had a sophisticated
understanding of what we now call photosynthesis. He specifically
cited the work of the Dutch scientist Jan Ingenhousz, who had
published a paper describing the process in 1779.[8] Hutton gave a
place to phlogiston in his own explanation and may have been
influenced by the work of Joseph Priestley, who was sometimes a
collaborator and sometimes a rival of Ingenhousz. Like Hutton,
Priestley clung to the phlogiston theory despite the work of
Lavoisier. He was a member of the Lunar Society, and it is possible
Hutton may have met him during his stay in Birmingham.

The manuscript, however, also included a discussion of the
economy of the West Indies and a description of the ownership of

African slaves which appears to be in direct contrast to the unequivocal condemnation of slavery which he expressed in his *Investigation
of the Principles of Knowledge* (see Ch. 19).

Happy African that now cultivates those fields! – your task is
moderate, and your maintenance is sure; you rear a family in
peace, and you see your children perhaps as happy as any
thoughtless people upon the Earth.

What tho' that African does not consume the sugar which he
cultivates; Is his table less supplyed [sic] with the necessaries of
life? And does he dance mightily with less animated joy, and
those who in the European assembly consume his coffee and
his sugar? He labours as every man must do, in order to aquire
[sic] his livelihood; he labours to his contentment, in the sun
which never offends his woolly head; and when his easy task is
over he then enjoys that pleasure (of idleness) which only can
be purchased by labour.

No man more interested in the health and population of the
labourers which he employs, than our W[est] I[ndian] Planters;
for, in this respect their interests are inseparable. But the feeding of that people, in great measure comes from other climates;
the corn that grows in England, and the herrings that are cured
in the coast of Scotland, are eaten in the West Indies, and are
exchanged for the produce of the cane. Thus two very different climates are made to cooperate for the benefits of man; and
the fittest climate perhaps on earth for maintaining immense
population or for growing the necessaries of life, is made to
feed its industrious inhabitants by growing nothing but the
luxuries, which are to be consumed in another country.[9]

Hutton had never visited the West Indies and appears to have
accepted the propaganda of slave owners. Nowhere in the section
does he use the word 'slave' and he appears to believe that Africans
worked the plantations voluntarily, rather than under violent duress.
There were frequent slave revolts in the Caribbean during the 18th
century, but it would be another 40 years before slaves were freed
in British colonies. It is perhaps a reflection of his failing health that
he did not question how and why they crossed the Atlantic.

Hutton's sentiments were not unusual. In 1791 the writer and lawyer James Boswell had published a poem entitled 'No Abolition of Slavery, or the Universal Empire of Love', making similar claims about the lives of slaves. It included the verse:

The cheerful gang!—the negroes see
Perform the task of industry:
Of food, clothes, cleanly lodging sure,
Each has his property secure;
Their wives and children protected,
In sickness they are not neglected

★ ★ ★

Playfair tells us that Hutton's condition deteriorated during the winter of 1796–7, he found difficulty in eating, and became 'extremely emaciated' and progressively weaker. He was in acute pain, but his mind was still sharp and clear, we are told, and he continued to read avidly. During the winter he received the fourth volume of Saussure's *Voyages dans les Alpes* and devoured it. Fittingly, Playfair comments, the last work of one eminent geologist became the final study of another. The end, when it came, was sudden and peaceful:

On Saturday 26th March he suffered a good deal of pain; but, nevertheless, employed himself in writing, and particularly in noting down his remarks on some attempts which were then making towards a new mineralogical nomenclature. In the evening he was seized with a shivering, and his uneasiness continuing to increase, he sent for his friend Mr Russell, who attended him as his surgeon. Before he could possibly arrive, all medical assistance was in vain: Dr Hutton had just strength left to stretch out his hand to him and immediately expired.[10]

Robison gave the news to Watt:

Dr Black's health is better than when I formerly spoke of it, but he feels very strongly the loss of his amiable and worthy friend and companion Dr Hutton. Dr Black has not had spirits enough to be able to see Dr Hutton these several months till within a few days of his decease. He left us without a struggle in less than half a minute after speaking with the utmost clearness. He was busy with another large volume and had engaged the engraver to come and get his orders the day after that on which he died.[11]

Hutton's mystery continued after his demise. The *Edinburgh Advertiser* reported his death in a line and a half, less space than it gave to the unfortunate three-year-old girl run over by a cart in Aberdeen or the young blacksmith accidentally shot dead in Dumfriesshire.[12] Playfair does not mention a funeral and there are no contemporary accounts. We don't know if his friends and surviving sister accompanied him to his grave. According to the Old Parish Register, held in the National Records of Scotland, he was buried the day after his death. We do not even know for certain where he was laid to rest. The place of burial is given in the register as Greyfriar's Kirkyard – 'Mr Gray's ground, close to N.E. Chas Keiss headstone'.[13] The exact location is unclear from this description.

Tradition has it that Hutton was not buried, but interred alongside his mother and other relatives in the Balfour family vault in an area of Greyfriars Kirkyard known as the Covenanters' prison (400 religious dissenters were confined there after the Battle of Bothwell Bridge in 1679). But his presence in the vault – if indeed he was there at all – was unmarked until 1947, when a plaque was erected by a group of geologists to commemorate 150 years since Hutton's death.[14]

Hutton was meticulous in cataloguing his mineral collection, but his personal papers were in a mess. Black complained to Watt that Isabella, being unaccustomed to business, was relying on him, but sorting out the financial affairs would cost him time and trouble.[15] There was no will, but Black engaged John Bell, the son of Hutton's long-time lawyer, business adviser and friend, to act for Isabella in obtaining a 'retour' – a legal document transferring all his assets and possessions to her, as apparently his only living heir.[16] It did not detail what the assets were, but from Isabella's will, published after she died in 1818, we can see that Hutton, although not rich, was

certainly very comfortably off at the time of his death.[17] Isabella
inherited a number of properties, including the house in Gosforth
Close, where she and James had been born and grew up. She was
receiving rents from 14 tenants of properties in Edinburgh and
from the farmers of Slighhouses and Monynut and she was owed
interest payments from a number of loans she had made. These
included £500 to her nephew Andrew Balfour and £600 to
Edington Smeeton Hutton, one of James Hutton's grandsons, but
also £6,500 to John Bell and his brother Archibald, who had probably invested it on her behalf. Her moveable and personal estate
(not including land and property) was valued at £1,500.

Surprisingly, Isabella's will disclosed that, in 1799, two years after
her brother's death, she had lent John Davie, his former business
partner, the large sum of £4,990, which was outstanding at the
time of Davie's death, together with penalties and interest adding
up to a further £998. The security for this loan was Davie's estates
at Gavieside in West Lothian. She also owned land near Lasswade
in Midlothian, where a quarry had been dug and a paper mill built.
Hutton may have left cash in a deposit account in one of the private
banks (probably Sir William Forbes & Co; there is no record of an
account in his name in any of the three public banks then in existence – Bank of Scotland, Royal Bank of Scotland and the British
Linen Company). At her death Isabella had nearly £60 in her
account with her solicitor. Hutton's income from the partnership
with Davie probably ended with his death.

We don't know how long after Hutton's death his son arrived in
Edinburgh or how he had heard about it. There were no national
newspapers and Hutton's passing merited only brief notices in the
Edinburgh papers, so probably went unreported in Cumbria where
he was living. Was he informed by letter either that his father had
died, or was gravely ill? Was it his father wanting to see his son again
before he died who wrote to him, or was it his aunt Isabella? If it
was her, she was clearly in on the secret. So was John Davie, who
had agreed to the bond the Hutton-Davie partnership had offered
to the younger Hutton 20 years previously.

Black told Watt that the younger Hutton had seven children –
four sons and three daughters. 'As the Doctor's property is considerable we hope to get a settlement made by Miss [Isabella] Hutton

in which something handsome will be done for Mr Hutton and his family.'[18] Isabella outlived her nephew by 16 years, so the will does not specify any provision for him and we do not know whether she transferred money to him and his family during his lifetime.

Black personally helped William Smeeton Hutton, the grandson who was apprenticed to a surgeon in Workington and 'much spoken of for his genius and parts', arranging for him to continue his medical studies in Edinburgh. When James Hutton junior died, William was left 'his grandfather's watch' in his father's will. A second son, Edington Smeeton Hutton, was initially helped by Black to find a job in a financial company in London, but when that did not work out Black asked Watt to find him a post in Boulton and Watt. Edington stayed with the company for some years, working as a steam engine installer. He died in 1831. A third son, Thomas Houston Hutton, joined the East India Company and in 1806, with financial help from Isabella, purchased a commission as ensign in the 77th regiment in Bombay. The oldest son, James Edington Smeeton Hutton, may also have joined the army, but we know little about him. A fifth boy, Francis Smeeton Hutton, had been born in 1795, but died four years later. James Hutton junior also had three daughters. Margaret, the oldest, was 20 when she accompanied her father to Edinburgh in 1797. She later married, as did her sister Jane, but Alice appears to have remained single.

After stipulating that her debts and funeral expenses be paid, Isabella made bequests to her surviving relatives. She left £2,500 to Andrew Balfour and £1,500 to each of his two sisters, Agnes and Margaret. The women also were left the contents of the house at St John's Hill, cutlery, china, bed- and table-linen. There is no mention of the house itself, which was later occupied by Andrew and his family.[19] Margaret had been a companion and nurse to Isabella and received a further £1,000. In a codicil to the will John Hutton Balfour, Andrew's oldest son, was also left £1,000. After donations of £50 to the Royal Infirmary of Edinburgh and the same amount to the orphans' hospital, she left an annuity to Alice Hutton, the widow of the younger James Hutton. The remainder of her estate, which included the two Berwickshire farms and the Edinburgh properties, were left to Hutton's grandchildren, with a stipulation that £1,000 should go to Edington Hutton, as the oldest grandson.

The money he had already borrowed from Isabella was to be deducted from this and he was to use some of the cash to buy a commission in the Indian army for his brother Thomas.

Isabella was anxious that women should have their independence. There was an express wish that the legacies which went to Margaret and Agnes Balfour and to the three Hutton granddaughters should go to them exclusively and not to their husbands, if they had married.

Chapter 21

Ages may be required

A MYSTERY SURROUNDS the fate of Hutton's work papers. According to Playfair: 'As he was indefatigable in study, and was in the habit of using his pen continually as an instrument of thought, he wrote a great deal, and has left behind him an incredible quantity of manuscript, though imperfect, and never intended for the press.'[1] What happened to this large volume of paper? Did it pass to Black or to Playfair, and if so what did they do with it? Did Hutton ask them to burn it, as he and Black had done with Adam Smith's papers?

Despite searches by many scholars, none of it appears to survive. In 1892 a James Hutton, describing himself as a 'distant kinsman' of Hutton wrote to Sir Archibald Geikie, then president of the Geological Society of London, praising an address he had given and adding that he wished he knew how his family's estate 'had passed into the hands of Mr Home of Milne Graden'. The Home family were large landowners in Berwickshire, and the writer of the letter was probably referring to the farms, but in the hope that any part of the Hutton estate which passed to them might include a cache of manuscripts, the researcher Jean Jones interviewed the then owner of Milne Graden in 2000, but failed to find any Hutton papers. The Milne Graden papers are now in the National Records of Scotland, but contain no reference to Hutton.[2]

We also do not know what happened to Hutton's library, nor his scientific instruments and apparatus. There is no mention of them in Isabella's will, when she lists the contents of the house in St John's Hill which were left to her Balfour cousins. Hutton's geological collection – his 'library of books written upon with God's own finger' – has also disappeared. We know that Isabella gave the collection to Joseph Black, who was anxious that it be preserved and made available to scholars and so passed it with a number of conditions to the Royal Society of Edinburgh, which appointed Playfair and Sir James Hall as trustees. But in 1821, after the death of Playfair and the retirement of Hall, there is a reference in the society's draft minutes to the appointment of Sir George Mackenzie and Dr William Brewster as custodians.[3] The society had nowhere to keep the collection and so seems to have transferred it to the university for its Natural History Museum, where it fell into the hands of Hutton's opponents.

In 1804 Robert Jameson became the professor of natural history at Edinburgh. He was a fierce critic of Hutton's theory and a supporter of Abraham Gottlob Werner, a hugely influential and experienced German mineralogist and teacher who believed the Earth had once been covered by a huge ocean. However, he did not subscribe to the Noah's Flood narrative and accepted that geological time had to be measured in millions of years, rather than the thousands of the biblical scholars. Werner's view that rock strata had all been laid down by a primal ocean became known as 'Neptunism'. In contrast, Hutton's assertion that subterranean unerupted rocks being pushed slowly upwards, 'Plutonism'.[4] The belief that volcanic activity played an important part in shaping the Earth was called 'Vulcanism'. The terms gained currency, although neither Werner nor Hutton ever used them.

Jameson had been an opponent of Hutton's theories even before he became a professor. In 1796, when a medical student aged 22, he read two pretentious papers to the Royal Medical Society of Edinburgh, objecting to Hutton's theory, the first on the origin of basalt, the second entitled 'Is the Huttonian Theory of the Earth consistent with fact?' Later, as a professor and the man in charge of the university museum he actively prevented access to Hutton's geological collection, did not display it and allowed it to be broken up and dissipated. It is impossible now to say that any mineral samples in the Royal Scottish Museum, which inherited the university museum's collection, definitely came from the Hutton collection.[5] Did Hutton's papers meet a similar fate at the hands of Jameson or some other opponent of Hutton's theory, or did Black destroy them following some verbal instruction from Hutton himself?

One manuscript whose detailed history we do know is that of *Elements of Agriculture*, which was sent by Playfair to the Edinburgh bookseller and publisher Archibald Constable. He arranged for Cadell & Davies to publish the book in partnership with John Murray. Playfair's biography of Hutton was to be included in the volume. If the book was published – and there is some evidence from 1854 that some copies were printed – it has since been lost. However, the original manuscript survived, found its way via the Edinburgh Geological Society and the Royal Scottish Geographical Society to the Royal Society of Edinburgh, which loaned it for safe-keeping to the National Library of Scotland, where it is until this day.[6]

The story of the two concluding sections of *Theory of the Earth*, on which Hutton was working when he died, is more tortuous. It would be more than a hundred years before any of volume 3 appeared in print and volume 4 has disappeared altogether – if indeed it ever existed. After Hutton's death Isabella passed the unfinished manuscripts to Playfair, perhaps hoping that he would complete the work and publish it. Instead he gave it to Lord John Webb Seymour, brother of the Duke of Somerset and a keen amateur geologist. After graduation from Oxford Webb Seymour had moved to Edinburgh, become a fellow and later a vice-president of the Royal Society of Edinburgh. He and Playfair became very close (Cockburn described them as 'husband and wife')[7] and had undertaken a number of field trips together, including to Glen Tilt.

Before his death in 1819, Seymour gave the manuscript to Leonard Horner, son of an Edinburgh merchant and another keen geologist. Horner kept it for over 25 years before handing it to the Geological Society of London for safekeeping in their library.[8] There it lay until Sir Archibald Geikie, director of the Geological Survey of Scotland, persuaded the society to publish it in 1899.[9] By that time the first three chapters had been lost, so that volume 3 starts with Chapter IV on page 139. There were no illustrations: Clerk's pictures remained missing until rediscovered in the archives of the Clerks of Penicuik and published in 1978,[10] but Geikie shouldered his camera and retraced Hutton's steps on Arran, taking photographs at the places he believed Hutton would have studied.

Interestingly, a review published shortly after the third volume appeared in print praised the work for showing Hutton as an accomplished field geologist, rather than the mere speculative theorist that some had accused him of being. Despite contemporary criticism of Hutton's ponderous prose style, the Victorian reviewer also lauded his ability to 'describe accurately and reason profoundly in the ordinary English tongue; and this is not the least charm in days when geological writing is apt to become a conglomerate of scientific jargon unintelligible to all but specialists'.[11]

Hostility to Hutton's work in the form of direct attacks from opponents like Kirwan did not end with his death. He had very few supporters outside his friends in Edinburgh and James Watt and Erasmus Darwin in the Midlands. Joseph Black, Hutton's most stalwart advocate, died of a heart attack two years after his friend. Cockburn described Black dying seated with a bowl of milk on his knee: 'which his ceasing to live, did not spill a drop'.[12] His passing left Playfair and Sir James Hall to defend and promote Hutton's theory. The Royal Society of Edinburgh was at the forefront of geological knowledge, with papers presented not only by Hutton, Black, Playfair and Hall, but also by Black's protégé Thomas Hope, who read a paper on the newly discovered element strontium.[13] Sir George Mackenzie, a friend of both Black and Playfair, lectured on the combustion of diamonds. Interest in geology built to such an extent that by 1811 geological papers were almost outnumbering all other scientific papers combined and led to the establishment of a geological committee.[14]

Less than a year after Hutton's death, Hall restarted his experiments to prove parts of Hutton's theory or at least refute some of the objections to it. Critics like Kirwan had claimed that Hutton was wrong in asserting that deep in the Earth rock was kept liquid by intense heat and flowed to the surface as lava during eruptions. Granite when heated, Kirwan claimed, turns to glass, whereas lava is obviously still rock. In March and June 1798 Hall read papers to the RSE describing how, with advice from Thomas Hope and others, he had heated samples of whinstone and then of lava in furnaces. He showed that the rate of cooling determined whether or not they became vitreous, and he had successfully turned stone to glass and back to stone.

Experiments of this kind were not entirely new. James Keir had described similar findings to the Royal Society of London in 1776 and others were attempting to melt granites and basalt.[15] Hall concluded that his results strongly confirmed Hutton's ideas and added: 'I flatter myself, likewise, that the experiments, independently of the general views of geology, are of some value, by accounting for the stony character of lavas, and thus enabling us to dispense with the various mystical suppositions which have of late perplexed the history of volcanic phenomena' – a clear reference to a paper by Kirwan on lava.[16]

In 1800, prompted by Hall's experiments and an earlier paper by him on granite which confirmed Hutton's observations,[17] Kirwan returned to the attack. In a presentation to the Royal Irish Academy he disputed Hall's conclusions, adding patronisingly: 'fanciful and groundless though the Huttonian theory seems to me to be, it may, like the researches for the philosopher's stone be highly useful by suggesting new experiments'. Grudgingly he acknowledged that Hall's work might have done a service to geology, but was unshakeable in his belief that they did not justify 'the many hypotheses heaped on each other by Dr Hutton'.[18]

Playfair's decision not to publish the remaining volumes of Hutton's final restatement of his theory was driven by a belief that the sheer length of the work and Hutton's difficult prose were barriers to the acceptance of his ideas. His solution was to produce a shorter, simplified summary himself, which he read to the Royal Society of Edinburgh during two meetings in 1799. According to one of his listeners it was 'a very distinct and luminous deduction from as powerful a train of arguments as ever was given in favour of a mere hypothesis'.[19] Later, with the addition of notes, explanations and some rebuttals of the criticisms of Jameson and Kirwan, he expanded the lectures into a book, *Illustrations to the Huttonian Theory of the Earth*, which appeared in 1802.

In the foreword he stated his purpose: to give Hutton's theory 'in a manner more popular and perspicuous than is done in his own writings. The obscurity of these has been often complained of and thence, no doubt, it has arisen, that so little attention has been paid to the ingenious and original speculations which they contain.'[20] Playfair rewrote and reordered Hutton's text and edited out many

of the digressions and unnecessary examples. He defended his right
to do so:

> Having been instructed by Dr Hutton himself in his theory of
> the earth; having lived in intimate friendship with that excel-
> lent man for several years, and almost in the daily habit of
> discussing the questions here treated of; I have had the best
> opportunity of understanding his views, and becoming
> acquainted with his peculiarities, whether of expression or of
> thought. In the other qualifications necessary for the illustra-
> tion of a system so extensive and various, I am abundantly
> sensible of my deficiency, and shall therefore with great defer-
> ence, and considerable anxiety, wait that decision from which
> there no appeal.

Playfair did not merely explain Hutton's arguments, he also
expanded some of the concepts, brought his own expertise as a
mathematician and physicist to bear on the theory and accepted
points where Hutton was either wrong, or had missed important
facts. Where Hutton had described 'fire' at the centre of the Earth
as one of the great agents of change – leaving himself open to the
unanswered questions, what fuels this fire and how does it persist
over aeons without oxygen – Playfair talked of 'heat'. He also introduced the 'principle of pressure' as an agent to modify the effects
of heat, again an advance on Hutton's original formulation.[21] To
explain how this heat could exist he called in aid no less an authority than Newton: '"Are not", says he, "the sun and fixed stars great
earths, vehemently hot, whose heat is conserved by the greatness of
the bodies, and the mutual action reaction between them, and the
light which they emit?"'[22]

In an attempt to justify Hutton's most misrepresented concept,
that the process of destruction and renewal of the Earth had no
perceptible beginning or end, Playfair quoted the French mathematicians La Grange and Laplace on planetary motion to show
that repeated processes of 'unlimited duration' did exist in the
natural order of the universe.[23] This drew on science which Hutton
could not have known about; Laplace's major book appeared only
after his death. To cite French sources was brave, even provocative.

Since the Revolution anti-French feeling had extended from politics and religion to science. John Robison, Playfair's predecessor as professor of natural philosophy at Edinburgh, had led attacks on French scientists including Laplace and Lavoisier for their supposed 'Jacobinism' and atheism.[24] Playfair sought to defend Hutton against charges of atheism by asserting that his view of the history of the earth was not in conflict with Genesis, which was concerned with the much more recent history of mankind on which 'geology is silent'.[25]

Playfair appeared to have modest ambitions for his recasting of the theory of the Earth, writing that although Hutton had established the basic principles 'on an immovable basis', the theory was imperfect:

> A work of such variety and extent cannot be carried to perfection by the efforts of an individual. Ages may be required to fill up the bold outline which Dr Hutton has traced with so masterly a hand; to detach the parts more completely from the general mass; to adjust the size and position of the subordinate members; and to give to the whole piece the exact proportion and true colouring of nature.[26]

Playfair's attempts to shore up Hutton's arguments still left some fundamental questions unanswered. A long appraisal of *Illustrations* in the *Edinburgh Review*, probably written by its editor Francis Jeffrey, focused on the importance of heat. Playfair may have skirted the objections to 'fire' but he still left unexplained the origin of this heat, how it lasted for so long and why it appeared to act in contradictory ways at different times. It was against the rules of philosophy, the review asserted, to invent causes in order to explain effects.

> If we were allowed to suppose an inexhaustible heat in a situation where our experience tells us that no heat could be either generated or maintained, it will not be easy to show why we should refuse to believe that a dragon eats up the moon in an eclipse, or that the tides are occasioned by the gills of a leviathan at the pole.[27]

Heat was the first of a series of detailed objections which the reviewer claimed had not been answered by Playfair. The *Review*, like Playfair himself, favoured the Whig political party and Playfair was a friend of Jeffrey. Nevertheless, the article ended after 15 pages praising him for his industry and diligence, but remaining unconvinced by Hutton's theory. 'It is a system built on postulates so bold and involving operations so prodigious, so capricious and so incapable of exemplification from actual experience, that we do not conceive it susceptible of any complete or satisfactory defence.'[28] Despite Hutton's lifetime of observation and field research, in the minds of most of his readers his theory was mere speculation.

A much longer critique of Hutton appeared at the end of 1802. *A Comparative View of the Huttonian and Neptunian Systems of Geology: In Answer to the Illustrations of the Huttonian Theory of the Earth, by Professor Playfair* was a 256-page book by John Murray, a doctor who lectured on chemistry and pharmacy. He also identified the source of heat as the major flaw in Hutton, but made detailed comparisons between Hutton and Werner's explanation for various geological phenomena. Although he allowed Hutton some credit, he came down firmly on the side of Neptunism as providing the more plausible explanations for the development of the Earth. 'To the Huttonian system belongs the praise of novelty, boldness of conception and extent of views,' Murray allowed '. . . and he has succeeded in drawing an outline which gratifies the imagination with the semblance of grandeur and design. But these are the only merits of the theory and they have certainly been much over-rated by the partiality of its defenders.'[29]

A more insidious attack was made the following year by the Rev. William Richardson. His paper purporting to disprove Hutton on basalts by examining samples supposedly found on the coast of Northern Ireland close to the Giant's Causeway was first read in his absence to the Royal Society of Edinburgh, where it was not well received. Two months later he appeared in person to read the same paper to the Royal Irish Academy in Dublin, but this time prefacing his geological arguments with loosely veiled accusations of atheism. He began by accusing Hutton of vanity:

Philosophers at all times seem to have been seized with a sort of rage for inventing and supporting theories, and for explaining the operations of nature and the phenomena she exhibits upon principles discovered by themselves; they seem to have considered it as humiliating to admit they were not privy to her secrets and that they were unable to explain the manner in which her various works were executed.

The philosopher Francis Bacon had warned against 'false doctrines and theories', but Richardson added: 'he did not foresee that at a future period in the hands of Antichristian Conspirators, they would be made instruments to support infidelity, concealed under the mask of mere physical opinions whose object was to shew that a much longer space of time is required for the formation of the universe than the history of the Creation, as delineated by Moses, leaves us room to suppose.' He then added, disingenuously, 'That Dr Hutton, in inventing the theory called by his name, had any such object in view, I by no means insinuate. It would be unfair to impute to any man motives he does not avow and illiberal to attack a gentleman who, having paid the debt of nature, can no longer defend himself.'[30]

After a further attack in 1804, Richardson was taken by Sir James Hall, Thomas Hope and David Brewster, a Presbyterian minister and self-taught scientist, to Salisbury Crags just behind Hutton's former home in St John's Hill. He was shown the famous junction of sandstone and basalt (now known as 'Hutton's Section') with its one piece of sandstone entirely surrounded by intrusive basalt, which Richardson had denied could exist. 'When Sir James had finished his lecture,' Brewster recalled years later, 'Richardson burst out into the strongest expressions of contemptuous surprise, that a theory of the earth should be founded on such small and trivial appearances!'[31] Clearly he was not to be persuaded by facts.

Even in the RSE, supporters of Hutton and acceptance of his theory was not total. In 1808 Robert Jameson led what was effectively a breakaway group to found the Wernerian Natural History Society, which as its name suggests favoured the Neptunism of Werner over the Plutonism of Hutton. Its honorary members were Werner himself, Sir Joseph Banks, president of the Royal Society of

London, and Richard Kirwan, president of the Royal Irish Academy. Its ordinary members included some surprising names: James Watt, Humphrey Davy, Robert Stevenson, the lighthouse engineer and grandfather of Robert Louis Stevenson, and Alexander Monro, who had followed his grandfather and father as a surgeon and professor of anatomy.[*]

The society was a serious scientific body and, although many of its members were also fellows of the RSE, it soon began publishing papers and appeared to be a rival to the RSE as the pre-eminent scientific academy of Scotland. The Wernerian Society was not entirely devoted to geology. Its first collection of 33 essays published in 1811 included many on zoology and botany as well as geology. Jameson was the most frequent contributor, but his papers were mostly concerned with practical geology, rather than any overt contradiction of Hutton's theory. In fact the only reference to Hutton in the first three years of the society's activities was made not by Jameson, but by James Ogilby, a doctor of medicine, who had examined a piece of greenstone picked up in the Lammermuir hills south of Edinburgh and disagreed with the assertion by Hutton and Playfair that it was a type of granite.[32]

Jameson had studied under Werner in Freiberg and set out his adherence to his theories in his three-volume book *The Wernerian theory of the Neptunian orgin of rocks*, which was published in stages from 1805 to 1808. Although the book described Werner's view of the origin of the world and adopted his classification of rocks, it was also intended as a textbook of the current state of geology. A refutation of Hutton's theory was only included in notes at the end. Jameson has been blamed for the failure of Hutton's ideas to make progress in the early part of the 19th century.[33] He was undoubtedly hostile to Hutton's theory, but his influence was not extensive and he does not seem to have gone out of his way to influence his students. His lectures were described as 'desperately dull, doleful' and Charles Darwin found them so uninspiring that he determined never to read a book on geology.[34] Jameson played a leading role in the Wernerian Society until his death in 1854. It did not long outlast him, closing in 1858.

[*] Tertius, the least distinguished of the Monro dynasty.

If Jameson was not the villain, why did Hutton's theory fail to find new supporters, even after Playfair's efforts? The geologist Derek Flinn believed that with the advance in experimental techniques, scientific opinion after 1800 had turned against grand theories and Hutton was regarded as a member of an old-fashioned school of 'philosophical geology'. The spirit of the age demanded fieldwork and evidence and this led many younger geologists to reject not just Hutton, but Werner too and others as mere theorists.[35] The Geological Society was founded in London in 1807 and committed itself to the discovery of fact, rather than the promotion of theories,[36] while the *Edinburgh Review* became increasingly dismissive of Wernerian theories as well as Huttonian ones.[37] The *Monthly Review* grew tired of the whole controversy: 'the discussion of the two rival systems of fire and water has been too frequently agitated to afford either instruction or amusement to our philosophical readers'.[38]

Playfair continued to champion Hutton in book reviews and articles. In 1816–17 he made an extensive tour of Europe to gather material for a new comprehensive book on geology, but his health deteriorated, and he died in 1819 without beginning the work. Sir James Hall continued his experiments, showing, among other things, how strata could fold and how sand could be turned into sandstone. He entered politics as MP for a constituency in Cornwall and published a book on Gothic architecture, but his last 20 years were also dogged by ill health and he died in Edinburgh in 1832.

Chapter 22

Hutton every day strikes me
with astonishment

CHARLES LYELL WAS born in 1797, the year that Hutton died. Like Playfair he was from Forfarshire, although that is where the similarity ended. Playfair was brought up in the manse in his father's parish, which he inherited when he became a minister in the Church of Scotland (a post he relinquished on entering academic life). Lyell's family made a fortune supplying the Royal Navy at Montrose and bought an estate at Kinnordy, near Kirriemuir. Lyell was born there, but when he was a few months old his family moved to the south of England, renting an estate near the New Forest where, as a growing boy, he became interested in the local natural history. Lyell's father wanted him to become a lawyer and he was called to the bar and worked for a while as a barrister, but geology became his consuming passion.

His interest was first sparked by reading his father's copy of *An Introduction to Geology* by Robert Bakewell, a self-taught geologist who supported himself by giving public lectures.[1] The book, an easy read written in simple language, was aimed at a general audience rather than academics, who Bakewell believed sought to obscure geology in unnecessarily complex terms. It was ignored by the fashionable reviews when it first appeared in 1813, but sold so well that he produced an enlarged second edition in 1815. It eventually reached five editions, was translated into German and had a large circulation in the US. Bakewell took swipes at de Luc and Werner for making wild generalisations from limited local observations and at Kirwan for his intellectual snobbery. Although he did not accept Hutton's theory in its entirety, he praised its general concept – or at least Playfair's restatement of it, which he described as 'luminous and eloquent'. He was dismissive of Werner's Neptunian theory: 'It is scarcely possible for the human mind to invent a system more repugnant to existing facts.'[2] He also valued Hutton's fieldwork:

> I am inclined to think that the part of Dr Hutton's theory which relates to the igneous origin of basaltic rocks is as well established as the nature of the subject will admit of; other parts of the system are much less satisfactory. Mr Werner and most if not all of his disciples who deny the igneous origin of basalt, have never visited active volcanoes and seem disposed to close their eyes upon their existence.[3]

In 1817 Lyell began studying classics at Oxford, but also enrolled for the Rev. William Buckland's lectures on mineralogy. Buckland was a serious researcher who had written the first account of a fossil dinosaur and won a medal for his work on a Yorkshire cave which he proved had been a prehistoric hyena den. But as an ordained minister he clung to the biblical narrative and sought ways of reconciling it with his observations. While still a student, Lyell began a series of field trips which – probably coincidently – echoed those of Hutton. He began in Yarmouth, where, like Hutton, he was fascinated by the way the river constantly reshaped the coastline.

The following year he went to Paris with his parents and visited Le Jardin des Plantes (renamed from Le Jardin du Roi during the Revolution) to see the fossils collected by the palaeontologist Georges Cuvier and read accounts of the geology of the Paris Basin, published by Cuvier and Alexandre Brongniart in 1811. On a later visit Lyell was to meet both men and their fellow countryman Constant Prévost, who believed that the distribution of fossils in strata could be understood by looking at modern lakes, estuaries and seas. Like Hutton he thought present processes could explain past geological events.[4]

Lyell also made studies close to his home on the south coast of England and, like Hutton, visited the Isle of Wight. In Scotland, at Kinnordy, he examined the beds of drained lochs to confirm for himself the findings of Cuvier and Brongniart. He went to Staffa to study the basalt columns in Fingal's Cave and visited Salisbury

Crags in Edinburgh. While still a student, Lyell became a member of the Geological Society of London and remained a member, fellow or office bearer for the rest of his life. Over a long career of fieldwork he read a total of 45 papers to the society, but was careful to ensure that they all related to observations he had made or evidence he had collected, never to speculation. This was despite the fact that in his own lifetime he became famous and remains so now for his theory. George Greenough, the society's first president, who remained active throughout the 1820s and 1830s, was fiercely sceptical of attempts at theorising and became a critic of Lyell's books. In 1835 a member reported: 'Mr Greenough made a speech of some length, in his usual spirit of scepticism, attacking the Lyellian theory, with a good deal of humour, but not much argument.'[5]

The years leading up to the publication of the first edition of Lyell's *Principles of Geology*, which appeared in three volumes from 1830 to 1833, were periods of intensive fieldwork, particularly in the volcanic regions of the Auvergne, Naples and Sicily, studying ancient volcanoes and the still active Vesuvius and Etna. The book announced its intention in its subtitle: 'an attempt to explain the former changes of the Earth's surface, by reference to causes now in operation'.[6] From the very beginning Lyell distanced himself from the 'catastrophists', who believed sudden floods or eruptions had shaped the Earth (whether these were part of the biblical narrative or predated it). He also dismissed Buffon's theory, that the Earth had begun as a hot molten mass and had been steadily cooling since. He believed that the processes which shaped the Earth in the past were the same ones shaping it today. In this he was going against not only Werner and his followers, but more modern researchers like Cuvier and his own teacher and mentor, Buckland. Very early he established his debt to Hutton:

Werner appears to have regarded geology as little other than a subordinate department of mineralogy, and Desmarest included it under the head of Physical Geography. But the identification of its objects with those of Cosmogony* has been the most common and serious source of confusion. The first who

---

* The study of the origin of the universe.

endeavoured to draw a clear line of demarcation between these distinct departments, was Hutton, who declared that geology was in no ways concerned with questions as to the origin of things. But his doctrine on this head was vehemently opposed at first, and although it has gradually gained ground, and will ultimately prevail, it is yet far from being established.[7]

Lyell treated Hutton's *Theory of the Earth* as a foundation on which he could build. Hutton, he believed, had not satisfactorily explained how sedimentary strata were raised up to form mountains and hills. Playfair and Hall had attempted to fill this gap by suggesting periodic convulsions, which disrupted and bent the strata. Lyell rejected this idea: geological change was long, slow and continuous; the processes which shaped the Earth in the past, including volcanic eruptions and erosion by weather and water were still going on today.[8]

Lyell's ideas were revolutionary in the same way that Hutton's had been a generation before. He was attacked, not only by Greenough, but in the presidential address to the Geological Society in 1831, when the Rev. Professor Adam Sedgewick accused Lyell of misunderstanding – and even violating – the basic tenets of geological science. Like Hutton in the RSE, although Lyell had supporters within the Geological Society, many members were against him.[9] However, Lyell won other followers. In 1834 the Royal Society of London gave him a medal and his book proved so popular that he produced four new editions by 1838.*

It was a measure of how public opinion was changing that, in contrast to the equivocal reception Playfair's *Illustrations* had received, the *Principles* was favourably noticed in the *Edinburgh Review* in 1839. The writer, W.H. Fitton, who was a friend of Lyell and a member of the Geological Society, described the book as one of the most popular ever published and the most valuable since Playfair's explication of Hutton. It had been advertised as a beginner's guide – the sort of book which, Fitton added patronisingly, 'could have been read with ease and satisfaction by a well-educated

---

* It eventually went to 11 editions in Lyell's lifetime, and he was working on a 12th when he died. It was also translated into French and German.

woman' – whereas it was a clear and condensed abstract of a serious work of science.[10] He had a criticism:

> We think that Mr Lyell has not, either in his preliminary history or in the body of his works, done quite sufficient justice to the claims of Dr Hutton – whom we conceive to be effectively the author of the *Theory of the Earth* now almost universally received.[11]

Lyell took the criticism to heart and in a letter of reply to Fitton confessed: 'I found it difficult to read and remember Hutton, and though I tried, I doubt whether I ever fairly read more than half his writings and skimmed the rest.' He had included quotations from *Illustrations* at the beginnings of his books and told Fitton: 'The mottos of my first two volumes were especially selected from Playfair's Huttonian Theory, because although I was brought round slowly, against some of my early prejudices, to adopt Playfair's doctrines to the full extent, I was desirous to acknowledge his and Hutton's priority.' At the end of the letter he added: 'as an admirer of Hutton all I could have wished is that your panegyric on Hutton had appeared as aiding and seconding my efforts, since I trust that no book has made the claims of Hutton better known on the Continent of late years than mine.'[12]

Lyell had rescued Hutton from possible obscurity and, although he modified parts of Hutton's theory, had reaffirmed its basic principles. On many issues, the Huttonians had won, says Dennis R. Dean, in his magisterial work *James Hutton and the History of Geology*. The Edinburgh Geological Society, founded in 1834, was strongly Huttonian and effectively superseded the Wernerian Society. At the British Association for the Advancement of Science meeting at Edinburgh, held in the same year, it became clear that real opposition to Hutton had dissipated. But Lyell, rather than Hutton, was now the centre of controversy and debate within geology. 'From now on,' adds Dean, 'Hutton and Playfair belonged primarily to the historians' rather than geologists.

However, this did not make them immune from attack and in 1868 a new onslaught came from a powerful adversary, the physicist Sir William Thomson (later Lord Kelvin.) In a paper titled 'On

Geological Time', read to the Glasgow Geological Society, he called for a 'great reform in geological speculation' and in particular singled out Playfair's restatement of Hutton's famous phrase, that in relation to life on Earth 'we discern neither a beginning nor an end' and in relation to the motion of the planets, 'we discover no mark either of the commencement or the termination of the present order'.

As far as Thomson was concerned, nothing could be further from the truth. The idea of the almost limitless time needed for gradual geological destruction and renewal envisaged by Hutton and Playfair was against the laws of thermodynamics. Using some ingenious reasoning involving the friction caused by tides, he asserted with a precision worthy of Bishop Ussher two centuries before, that the rotation of the Earth was slowing down by 22 seconds per century and calculated that a billion years ago it would have been spinning at 17 times the rate it currently was so that 'bodies would fly off at the equator'. He concluded with an almost Neptunist position: the Earth must have been 'all fluid not many million years ago'.[13]

Thomson also believed that the Earth and the sun were cooling and therefore the Earth must have been hotter in the past – and he produced another calculation to dismiss, at least to his own satisfaction, the concepts both of uniformity and of geological time on which Hutton and Playfair had relied:

> The earth has not gone on as at present for a thousand million years. Dynamical theory of the sun's heat renders it almost impossible that the earth's surface has been illuminated by the sun many times ten million years. And when finally we consider underground temperature we find ourselves driven to the conclusion in every way, that the existing state of things on the earth, life on the earth, all geological history showing continuity of life, must be limited within some such period of past time as one hundred million years.[14]

Thomson's argument cast doubt not only on prevailing geological thinking, it also overshadowed Charles Darwin's theory of evolution, since it also envisaged the huge expanse of time described by

Lyell. Thomson's prestige and the fact that he appeared to be arguing from sound scientific principles and laid out his calculations for
scrutiny, led Lyell, Darwin and others, including Darwin's great
defender Thomas Huxley, to modify their views. For thirty years
Thomson seemed unassailable, and the more geologists and biolo-
gists suggested faults in his reasoning, the more he became
entrenched in his position. By 1897, using a calculation based on
the meting point of rocks, he claimed that the Earth was probably
only 20 million years old.[15]

* * *

Lyell's difficulty in reading Hutton's original prose, rather than
Playfair's more accessible summary, did not trouble Andrew
Ramsay, a Scottish-born geologist who had been working for the
Geological Survey in Wales. In 1848 he was appointed professor
of geology at University College, London, and in preparation for
his inaugural lecture reread both Playfair and Hutton. He noted
in his diary: 'Stuck at Hutton's *Theory of the Earth* and Playfair's
Illustrations all day (Sunday), and before night read all, and
made a complete abstract of the latter ... Hutton every day
strikes me with astonishment. Lyell does not do him half justice.'[16]
Fitton was among his audience at the lecture and would have
approved.

Ramsay was born in Glasgow, but spent childhood holidays on
Arran, which sparked his interest in rocks and mineralogy. In 1840
a geological model of the island by him was exhibited at the British
Association in Glasgow and in 1841 he published a guide to its
geology. Armed with this book, the young Archibald Geikie made
his first trip to Arran ten years later. Like Hutton, Geikie had been
born in Edinburgh, attended the High School and Edinburgh
University and had been apprenticed to a lawyer – a profession
which failed to ignite his interest any more than it had gripped
Hutton. In his brief summer holidays from the law firm he was
offered the choice by his family of a trip to London or geologising
on the island. Without any hesitation he chose Arran, which was to
become a lifelong source of fascination.

Geikie abandoned the law and joined the British Geological
Survey as an assistant in 1855. He was with the organisation for

over 40 years, becoming director of the Scottish branch in 1867 and succeeding Ramsay as the UK director in 1881. He was also the first professor of geology and mineralogy at the University of Edinburgh. Besides a lifetime of fieldwork, he was a prolific author. In his first book, *The Story of a Boulder*, which appeared in 1858, he used the example of the 'Hutton Section' of Edinburgh's Salisbury Crags to illustrate the battle between followers of Hutton and disciples of Werner and pinning his own colours to the mast:

> Many a battle was fought in this locality and not a few of the trap dykes and hills possess to the geologist a classic interest from having been the examples whence some of the best established geological opinions were first deduced. The contest between the Huttonians and Wernerians terminated long ago in the acknowledged victory of the former. Hutton's doctrines are now recognised all over the world. It is interesting however to walk over the scenes of the warfare and mark the very rocks among which it raged and from the peculiarities of which it took its rise.[17]

Geikie was intrigued to know what had happened to the unpublished portions of Hutton's final edition of the *Theory of the Earth* and searched for many years without success until, in the summer of 1897, he discovered that the manuscript of the third part had been in the library of the Geological Society of London since Horner had deposited it 42 years before.[18] The writings he found were incomplete, with a first section missing. Nevertheless he believed that the six chapters, which included one describing the field trip to Arran, showed Hutton in a new light.

> He was reproached in his lifetime with being a mere theorist. It must be admitted that from none of his published writings could it be inferred that he possessed so remarkable an observing faculty as is revealed in his dissertation on Arran. This striking essay is a masterpiece of acute observation and luminous generalisation. Had it been published in his lifetime it would have placed him at once as high in the ranks of field-geologists as he admittedly stood among those of the

speculative writers of his time. It seems but a tardy act of justice to his fame that the merit of this practical side of his life-work should be now at last fully established.[19]

Geikie appealed to the president and council of the society to publish the work, offered to edit it himself and had the manuscript copied. His research indicated that explanatory notes would be necessary to make the meaning of certain passages clear, but a notable gap was the absence of Clerk's illustrations, which were not with the manuscript. To provide substitutes, Geikie returned to Arran, which was being surveyed by William Gunn, one of the Geological Survey staff. Together they identified the locations described by Hutton and Geikie took photographs or made sketches. He worked on the book for a year, adding an index to the new section and to the two published volumes which were unindexed. He went to some trouble to make the new volume compatible with Hutton's original, even choosing paper and a type-face which he thought might have been used in the 18th century. The book – Hutton's original two volumes, plus the incomplete third volume – appeared in February 1899:

> Of all the publications with which I have been connected, none ever gave me more unalloyed gratification than the preparation of this volume, more than a hundred years after the death of its illustrious author. To be permitted to render any service, how slight soever, to the memory of this notable member of the group of founders of geology was an honour and a privilege of which no geologist could fail to be proud.

It was Geikie who called Hutton 'the father of modern geology', a title which has persisted ever since – although it is doubtful that Hutton would have approved and he never described himself as a 'geologist'. Geikie was fond of bestowing paternity: in a lecture to Johns Hopkins University, Baltimore (later turned into a book, *The Founders of Geology*), he named William Smith (1769–1839) as 'the father of English geology', and William Maclure (1763–1840) as 'the father of American geology'. Aristotle (384–322 BC) he named as the 'true father of natural history'.[20]

Geikie was one of the geologists who modified their positions to accommodate Thomson's calculation of the age of the Earth at 100 million years, but he rebelled when the estimate was reduced to 20 million years.[21] Thomson, by this time ennobled as Lord Kelvin, was coming under attack from several quarters. His one-time assistant and protégé John Perry, having failed to persuade his former boss to modify his view, wrote to the editor of the leading scientific journal *Nature* pointing out flaws in Thomson's assumptions about the conductivity of the Earth's core. His conclusion was that the Earth could be as much as 121 times older than Thomson believed[22] – hundreds of millions, if not billions of years old. A second blow came with the discovery of radioactivity, which provided a new source of heat for both the Earth and the sun. Ernest Rutherford, one of the pioneers of nuclear physics, addressed the Royal Institution in London in 1904:

I came into the room, which was half dark, and presently spotted Lord Kelvin in the audience and realised that I was in for trouble at the last part of the speech dealing with the age of the earth, where my views conflicted with his. To my relief he fell fast asleep but as I came to the important point, I saw the old bird sit up, open an eye and cock a baleful glance at me! Then sudden inspiration came, and I said Lord Kelvin had limited the age of the earth, provided no new source of heat was discovered. That prophetic utterance refers to what we are now considering tonight, radium! Behold! The old boy beamed at me.[23]

Chapter 23

What Newton achieved . . .

GEIKIE HOPED THAT the republication of Hutton's *magnum opus* would lead to 'renewed attention to his immortal work, which must ever remain one of the great landmarks in the onward march of science'.[1] But there seems to have been little new interest in Hutton's life, or examinations of his theory until 1947, when the Royal Society of Edinburgh organised a conference to commemorate the 150th anniversary of his death.[2] In his address, Sergei Tomkeieff, then a lecturer (later professor) at Newcastle University, compared Hutton to Newton. 'It is of little profit to argue as to which was the greater mind. What Newton achieved in the field of astronomy and mathematics, Hutton achieved in the field of geology.' Geikie had been keen to establish Hutton as a diligent field geologist, but Tomkeieff set him in the context of his time, an Enlightenment figure and first and foremost a philosopher: 'It was he who provided geology with a theory: and here we must take theory, not as something remote and opposed to fact, but in its original meaning in Greek, namely comprehension.'[3]

No other speaker contradicted Tomkeieff. In the 50 years since Geikie's rediscovery of Hutton, his contribution to our understanding of how the Earth had evolved had been accepted and the man himself hailed as one of the leading figures of the Scottish Enlightenment. Hutton's ideas had provided the core of modern geology, Tomkeieff added, although at first they had been imperfectly understood. It took Lyell and the work of men 'to whom the very name of Hutton may well even be completely unknown', to make the importance of Hutton's work clear. Playfair, although trying to do Hutton a service, had failed: 'Consciously or unconsciously Playfair presented Hutton's theory in such a way as to conform to the prevailing scientific tradition of his own time – the tradition of the departmentalised mind – and such a treatment could not have permitted him, even had he been capable of it, to sound the depths of the underlying philosophical foundations of Hutton's theory.'

Alongside Hutton, Tomkeieff credited Werner, who had produced a positional scheme of stratified rocks, and William Smith, the later English geologist who had discovered independently of Cuvier and Brongniart in France, that fossils could indicate the relative age of strata. These three had built the framework for modern observational geology.

> But science does not consist only of methods. It needs comprehension (theory) and this comprehension was provided by Hutton and Hutton alone. This is where the greatness and the genius of Hutton lie. Without him geology would certainly not have developed when and as it did.[4]

Hutton had faults which put obstacles in the way of the acceptance of his theory. One was his poor written communication, but the other was taking too far what was later known as his 'Plutonism' – the idea that all sedimentary rocks had been consolidated by the action of the internal heat of the Earth, including flints, septarian nodules and even coal. Yet Hutton was:

> A typical child of the age of reason, unbiassed by dogmatic prejudices, a free-thinker and a deist at heart, he gathered up all the leading ideological threads of his age and spun them into a masterly synthesis. This does not mean, however, that he was a mere compiler. His synthesis was of his own making and much of his material was also his own. Hutton was an extraordinarily acute and careful observer, but he was an even better reasoner.[5]

Hutton's reasoning was examined again in 1997 on the bicentenary of his death, by the French historian of geology Gabriel Gohau. Hutton, Gohau claimed, had not been as diligent a field geologist as Dolomieu or Saussure (whom he quoted in his books), but they had accumulated knowledge without being able or willing to put it in order. Hutton not only built a theory consistent with the evidence he had accumulated up until that time, but one which was able to predict what he would find later if he looked – the disrupted strata which he discovered on Arran, at Jedburgh and at Siccar Point.[6] Although disputing Geikie's claim that Hutton alone had founded modern geology, Gohau saw him as a revolutionary who had turned existing theory upside down: strata laid down at the bottom of the oceans became in Hutton's mind elevated high enough to form mountains, granite previously believed to be the 'primal rock' formed at the creation of the Earth had been shown by Hutton to be younger than some of the strata which lay on top of it. Again, he had been able to deduce this before going to find proof for it at Glen Tilt.

The strength of Hutton's theory, according to Gohau, was not only that it was consistent with the facts as Hutton knew them, but that it was able to accommodate later discoveries such as plate tectonics and continental drift. Quoting François Ellenberger in his *History of Geology*, Gohau says that a modern geologist browsing 18th-century books, would feel lost in a foreign universe, but reading the writings of the decade 1830–1840 [after Hutton and Lyell] would not feel disoriented, but easily recognise the modern discipline.

Other aspects of Hutton's theory have also remained intact. We can still argue over whether time is infinite, as Hutton postulated, but he was not interested in whether time had a beginning or not, he sought only to show that time was not a constraint on his argument. He was careful not to put a figure on the antiquity of the Earth, only to stress that the processes he was describing would take an unimaginably long time. Radiometric age dating has put the age of the Earth at more than 4.5 billion years – a figure several orders of magnitude larger than even the largest estimates of Hutton's critics. To be able to visualise a process which would take hundreds of millions or even billions of years made him unique among his contemporaries. He alone was able, in Playfair's words, to look into the abyss of time.

Hutton also had an ability to read the geological features before him and to imagine the events which caused them. He was not the first to find or describe 'unconformity', but its meaning had eluded his predecessors. As long before as 1669 Nicolaus Steno had made stylised sketches of disrupted strata, and they had been described by the English geologist John Strachey in 1719 and by the French mineralogist Jean Etienne Guettard in 1766. But only Hutton had understood their significance in the cycle of renewal and in providing convincing evidence of his theory.[7]

Another of Hutton's characteristics was his single-mindedness: he would not be drawn into fruitless disputes which distracted from his central argument. He was right about the internal heat of the Earth (although he was unwise to call it 'fire', corrected by Playfair to 'heat'). But he refused to be drawn into arguments about what caused the heat, what made it to all practical purposes eternal and why, over time, the whole of the Earth was not brought to the same uniform temperature. We now know much more about the heat of the Earth (partly caused by the radioactive decay of isotopes in the mantle and crust, and partly from the primordial heat left over from the formation of Earth) and we know that it has an effect on the geology of the surface, although in a more complex way than Hutton realised.

He was right about the gradual rising of land and he knew that volcanic eruptions could throw up new mountains, but he knew nothing of the much more important movement of tectonic plates. He was wrong in believing that heat played a crucial role in the formation of strata. We now know that compression alone is sufficient. Yet these more recent discoveries have reinforced rather than contradicted the concept of the self-renewing Earth machine he described.

Dennis R. Dean, in *James Hutton and the History of Geology*, identified ten ways in which Hutton contributed to our understanding of how the Earth was – and continues to be – formed:

1. The uniformity of natural laws throughout geological time ('uniformitarianism').
2. The destructive power of water.
3. The constructive power of subterranean heat.
4. The excavation of valleys by streams.
5. The igneous origin of dykes (a type of later vertical rock between older layers of rock).

6. The importance of compression.
7. The former fluidity and common origin of igneous rocks (later called 'Plutonic', although never by Hutton).
8. The uniqueness of subterranean processes.
9. The essential balance of creation and destruction.
10. The utility of geological processes for sustaining life.[8]

The biographies which have followed Playfair's first account of Hutton's life have all been by geologists or have concentrated on his geology. This is not surprising since his major contribution to our understanding is his *Theory of the Earth*. It was an elegant statement of the way in which the surface of the globe was formed and continues to be changed – even if it was hidden in Hutton's over-long and often awkward prose, in stark contrast to his simple verbal communication. Yet Hutton never described himself as a geologist (the term was not in wide use in his lifetime) or as a fossil collector, although he had many. His interests were wide. He trained as a lawyer's clerk and as a doctor. He was at one time a farmer, at another a chemical engineer and at another a civil engineer. He published an essay on language.

He described himself as a natural philosopher,[9] a searcher after truth. His longest published work was not his *Theory*, but a metaphysical treatise, which was greeted with almost complete silence on its publication and has been largely ignored since. His dissertations on heat, light and fire, on the nature and causes of rain, his ideas on the forces at work when matter expands and contracts, have largely been overlooked. He identified infrared light several years before Herschel published his 'discovery' and described survival of the fittest in animals and plants decades before Charles Darwin.

If Hutton is popularly known at all it is as the talented amateur who was intrigued by watching soil being eroded on his farm, mulled it over in his mind for 30 years and produced a comprehensive geological theory. He has been described as 'the man who found time',[10] and the person who estimated the 'true age of the earth'.[11] But the image of Hutton as the simple farmer developing his sophisticated theory of the Earth while managing the soil is as much romantic nonsense as Isaac Newton formulating his theory of gravity after an apple fell on his head, or James Watt inventing the steam engine after watching a kettle boil. By the time he was farming he already had a thorough theoretical understanding of the current state of geological knowledge and ideas. It also undervalues Hutton as a serious scientist, keen not only to make observations and conduct experiments of his own, but to learn from the discoveries and the mistakes of others.

Like Newton, he could see far because he stood on the shoulders of giants. He was aware of the currents of thinking about the age of the Earth and was not the first to challenge the biblical orthodoxy which put it at around 6,000 years. In fact, his was not the first theory of the Earth: there was a spate of books and theses published in the years leading up to Hutton's initial presentation and some of them were remarkably close to his final scheme. He did not 'find' time – although he did promote the difficult concept of infinite time – nor did he attempt to estimate the age of the Earth, although he was sometimes accused of doing so. His most famous phrase, 'we find no vestige of a beginning, no prospect of an end', did not describe an eternal Earth, but, as he clearly stated, the result of his 'present inquiry'.

Hutton knew the limitations of his theory and did not claim that he had discovered how the world began, or how it would end. His contribution to our understanding of the mechanisms which shape our environment is no less important for that. The tercentenary in 2026 of Hutton's birth, will also mark 250 years since the publication of *The Wealth of Nations* by his friend Adam Smith and 250 years since the death of the philosopher David Hume. His achievements are no less than theirs and he should take his place among the polymaths of the Scottish Enlightenment. In this book I have tried to give him that place.

Hutton's timeline

1726   Hutton is born in Edinburgh, probably in Gosforth Close, Old Town

1726–40   Edinburgh: educated at the High School

1740–3   Attends Edinburgh University to study humanities

1743   Begins 3-year legal apprenticeship, but only completes 2 years

1745   Attends Edinburgh University to study medicine, and signs a 5-year apprenticeship with Dr Young but only completes 3 years of this

1747–9   Paris: studies anatomy and chemistry at Le Jardin du Roi

1749   Leiden: presents his medical dissertation and graduates

1749–50   London, for several months, then returns to Edinburgh

1752–4   Norfolk and Suffolk: studies agriculture

Journeys to Isle of Wight, Corf Castle, Weymouth, Portland, Bath, Oxford, Catton, Cambridgeshire, Lincolnshire, Derbyshire, Yorkshire and Northumberland observing geological features

1754   Rotterdam, Netherlands, Belgium and northern France, studies agriculture and horticulture. Returns to Edinburgh via London

1754–64   Farms at Slighhouses, Berwickshire

1754   Visits Eyemouth for a party (the 'Feast of Baal')

1757   Visits Leadhills, South Lanarkshire, to survey the land for George Clerk

1764   Visits Crieff, Perthsire, Dalwhinnie, Fort Augustus, Inverness, Easter Ross (including Balnagowan, on the Cromarty Firth), Caithness, Portsoy, Banff and

	Aberdeen with George Clerk, as Commissioner for Forfeited Estates
1765–6	Starts sal ammoniac factory in Newington, Edinburgh, with John Davie
1767–74	Joins management board of the Forth & Clyde Canal, makes many trips across central Scotland surveying ground and advising on geological conditions
1770	Builds house in St John's Hill, Edinburgh
1774	Accompanies James Watt on journey from Edinburgh to Birmingham, via Cheshire salt mines
1774	Stays with Erasmus Darwin in Lichfield, visits Derbyshire to study geology and inspect work on the Trent and Mersey Canal
1774	Journeys on horseback via Dudley, Warwickshire, Bridgnorth, Shropshire, Wales, Wiltshire, Somerset to Bath and returns to Birmingham
1774	Returns to Edinburgh via Shropshire, climbing the Wrekin, north Wales, Anglesey, Manchester, Rochdale and Buxton
1774	Perthshire – 'at the foot of the Grampians' – to stay with 'Madam Young'
1785	Blair Atholl, Glen Tilt
1786	Loch Doon, Ayrshire, Galloway
1787	Arran: Glen Shant Hill, Goat Fell, Lochranza
1787	Jedburgh
1788	Dunglass, Siccar Point, Berwickshire
1788	Isle of Man
1797	Hutton dies in Edinburgh and is buried in Greyfriars Kirkyard

Glossary of geological terms

Basalt: igneous rock, fine-grained, almost black in colour, formed from a fairly runny type of lava.

Calcite: a common mineral which fizzes with dilute acid and is the main component of limestone. Chemically known as calcium carbonate.

Coal: sedimentary rock made up of compressed and carbonised plant material.

Continental drift: an early theory of the movement of continents over time. Now replaced by plate tectonics.

Culm: term used for fine-grained waste from anthracite coal, used as a poor-quality fuel.

Dyke: an intrusion of igneous rock, formed when magma flows toward the surface through cracks or faults, cutting through the rock layers.

Erosion: breakdown and removal of rock material by flowing water, wind or moving ice.

Expansion: a growth in volume which happens when most solids get hotter, but also when water freezes to form ice.

Explosive eruption: a volcanic eruption where viscous, gas-rich magmas burst out as explosions of pyroclastic material ('bombs', pumice and 'ash').

Extrusive (igneous): rocks formed by eruptions at the Earth's surface (see Volcanic).

Feldspar: a common mineral in igneous and some metamorphic rocks.

Fossil: any trace of past life preserved in a rock (includes animal tracks and burrows as well as shells, skeletons and impressions of soft flesh).

Gneiss: coarse-grained, metamorphic rock often showing a 'banded' texture due to separation of pale and dark-coloured minerals.

Granite: light grey or pinkish, coarse-grained, igneous rock, cooled slowly in large intrusions (same chemical composition as pumice).

Igneous, igneous rock: formed from magma, either erupted from a volcano or cooled below ground in an intrusion (Latin *ignis* = fire).

Lava: molten rock (e.g. basalt) erupted from a volcano. See also magma (which is not quite the same thing!).

Limestone: sedimentary rock composed largely of calcium carbonate (fizzes with dilute acid), usually formed from remains of living organisms.

Magma: molten rock with dissolved volcanic gases, beneath the Earth's surface.

Marble: metamorphic rock formed from limestone – also fizzes with dilute acid.

Oxidation: a weathering process where iron-rich minerals (usually) react with air to form rusty-brown oxides.

Plate, plate tectonics: plate tectonics describes the slow motion of rigid 'plates' of the lithosphere due to movement (convection) of the mantle beneath.

Sand: sediment particles from 0.1 to 2mm diameter. Most sand grains are made of quartz, a very hard and chemically resistant mineral.

Sandstone: medium-grained, sedimentary rock composed mainly of sand grains that have been cemented together.

Schist: metamorphic rock which has a shiny, foliated, medium-grained texture, often containing much mica.

Sediment: material deposited by water, wind or ice. Includes pebbles, sand, mud, organic remains (e.g. shells) and salts left by evaporation.

Sedimentary rock: any rock made up of sediment grains. Examples: mudstone, sandstone, limestone, rock salt, coal.

Septarian nodules (septarian concretions): the result of volcanic activity when mud and organic matter are trapped and sealed by pressure and desiccation (intense drying) in a cemented state.

Shale: Mudstone that has been compressed to form a fine-grained, flaky, dark-coloured sedimentary rock.

Sill: an intrusion of igneous rock along the bedding planes of the surrounding rock layers (rather than cutting through them – see Dyke).

Slate: metamorphic rock formed by compression and heating of mudstone. Much used for roofing as it splits easily into sheets.

Strata: layers of rock formed by deposition of sediment (and sometimes lava and pyroclastic material).

Unconformity: a boundary where one set of rock layers cuts across another, representing a gap in geological time when rocks were worn away.

Uplift: when a region is literally lifted up as the crust is squeezed by tectonic forces or, sometimes, following the melting of ice-sheets.

Volcano, volcanic: cone-shaped (sometimes!) mountain formed by eruptions of lava and/or pyroclastics. Volcanic means 'from a volcano'.

Volcanic ash: fragments of rock and pumice thrown out of volcanoes by explosive eruptions. The finest particles are carried long distances by winds.

Volcanic gases: gases, like water vapour and carbon dioxide, are dissolved in magma below ground, but are released at the surface as pressure drops.

Weathering: slow breakdown of rock at the Earth's surface, due to climatic and biological processes.

Glossary of scientific terms

Phlogiston and **Caloric** (called 'calorique' by Hutton): these were colourless, odourless, weightless substances thought to be present in some physical and chemical reactions.

The existence of phlogiston, proposed at the end of the 17th century and developed by scientists over the next 70 years, was thought to be given off when a substance burned. The theory was discredited by the French chemist Antoine-Laurent Lavoisier, who proposed that oxygen was the vital element in combustion. Hutton was aware of Lavoisier's work and experiments by the English chemist Henry Cavendish and admired their experimental methods, but he did not think they offered a complete explanation and so clung to phlogiston until a more comprehensive theory was offered.

It is a mark of Hutton's single-mindedness and rigorous reasoning that he supported the phlogiston theory even after his close friends and collaborators Joseph Black and Sir James Hall had abandoned it.

Caloric was a substance which was thought to flow from hotter bodies to colder bodies and could pass through pores in solids and liquids. The expansion of materials when they are heated was thought to be because of the addition of caloric and the reduction in volume on cooling was because caloric was given off. Even Lavoisier accepted caloric, and the theory was not abandoned until the 19th century. But Hutton rejected it in favour of other explanations, such as his own characterisation of heat and light as forces, rather than substances and Joseph Black's concepts of Latent and Specific heat.

Latent heat: Joseph Black introduced the term in 1762 to describe the heat added or subtracted from a substance when its state is changed but its temperature remains the same. Examples include when ice at zero degrees melts to water at the same temperature, where heat is absorbed, or vaporisation, where heat is given off. Black derived the name from the Latin *latere*, meaning to lie hidden.

Specific heat: Black noticed that the same masses of different substances needed different amounts of heat to raise them by the same temperature. He thus proposed that different substances required amounts of heat which were 'specific' to them. He gave the example of water, which needed less heat to raise its temperature by a certain amount than the same volume of quicksilver (mercury) to raise its temperature by the same amount.

Sources

Archieven van Senaat en Faculteiten, University of Leiden
Archives de l'Académie d'Agriculture de France
Bibliothèque Nationale de France (BNP)
British Library (BL)
British Postal Service Appointment Books
London Metropolitan Archives
Minutes of the town council of Edinburgh (TCM)
National Archives (NA)
National Library of Scotland (NLS)
National Museums of Scotland (NMS)
National Records of Scotland (NRS)
Roll book of the Merchant Company of Edinburgh
Transactions of the Royal Irish Academy
Transactions of the Royal Society of Edinburgh
Transactions of the Royal Society, London
University of Edinburgh archives and research collections (EU CRC)
University of Oklahoma Libraries, History of Geology Archives, Torrens,
    Hugh S: Collection, Series 1: Hugh S. Torrens Papers

BOOKS AND PAPERS BY JAMES HUTTON

Hutton, James, 1777, *Considerations on the Nature, Quality and Distinctions of Coal and Culm*, C. Elliot, Edinburgh, and Richardson and Urquhart, London
Hutton, James, 1788, 'Theory of the Earth; or an investigation of the laws observable in the composition, dissolution and restoration of land upon

the globe', *Transactions of the Royal Society of Edinburgh*, Vol. 1, pp. 209–303

Hutton, James, 1792, *Dissertations on Different Subjects in Natural Philosophy*, Edinburgh, Strahan and Cadell

Hutton, James, 1794 (1), *A Dissertation upon the Philosophy of Light, Heat, and Fire*, Cadell & Davies

Hutton, James, 1794 (2), *An Investigation of the Principles of Knowledge, and of the Progress of Reason, from Sense to Science and Philosophy*, 3 vols, Strahan & Cadell

Hutton, James, 1795, *Theory of the Earth with proofs and Illustrations*, Cadell & Davies, London, William Creech, Edinburgh

Hutton, James, 1795, *Vol. I, Theory of the earth with proofs and Illustrations*, Project Gutenberg, www.gutenberg.net

Hutton, James, 1795, *Vol. II, Theory of the earth with proofs and Illustrations*, Cadell, Davies and William Creech, Edinburgh. Reprinted London Geographical Society, 1899

Hutton, James, 1899, *Theory of the earth, with proofs and illustrations. Vol. III*, Archibald Geikie (ed.), Geological Society

Hutton, James, *Elements of Agriculture*, unpublished, National Library of Scotland

ARTICLES, THESES AND DISSERTATIONS

Allchin, Douglas, 1994, 'James Hutton and phlogiston', *Annals of Science*, 51:6, pp. 615–35, DOI: 10.1080/00033799400200461

Brackett, W.O., 1935, *John Witherspoon: His Scottish Ministry*, PhD thesis, University of Edinburgh

Brown S.W., 2008, *Smellie, William (1740–1795)*, Oxford Dictionary of National Biography. https://www.oxforddnb.com/view/10.1093/ref:odnb /9780198614128.001.0001/odnb-9780198614128-e-25753?rskey= 7uKWVQ&result=1

Clow, Archibald & Clow, Nan, 1945, 'Vitriol in the industrial revolution', *Economic History Review*, Vol. 15, No. 1/2 (1945), pp. 44–55

Clow, Archibald & Clow, Nan, 1947, 'Dr James Hutton and the manufacture of Sal Ammoniac', *Nature*, 29 March 1947, pp. 425–7

Davies, G.L., 1967, 'George Hoggart Toulmin and the Huttonian theory of the earth', *Bulletin of the Geological Society of America*, 78, pp. 121–4

Dean, Dennis R., 1975, 'James Hutton on Religion and Geology: the unpublished preface to his Theory of the Earth (1788)', *Annals of Science*, 32:3, pp. 187–93, DOI: 10.1080/00033797500200241

Dean, Dennis R., 1981, 'The age of the earth controversy: Beginnings to Hutton', *Annals of Science*, 38:4, pp. 435–56, DOI: 10.1080/00033798100200311

Dingwall, Helen, 1989, *The Social and Economic Structure of Edinburgh in the late 17th century*, Vols I & II, PhD thesis, University of Edinburgh

Donovan, Arthur & Prentiss, Joseph, 'James Hutton's Medical Dissertation', *Transactions of the American Philosophical Society*, Vol. 70, No. 6 (1980), pp. 3–57

Donovan, Arthur, 1978, 'James Hutton, Joseph Black and the Chemical Theory of Heat', *Ambix*, 25:3, pp. 176–90, DOI: 10.1179/amb.1978.25.3.176

Duckham, Baron F., 1969, 'Serfdom In Eighteenth Century Scotland', *History* 54, no. 181, pp. 178–97, http://www.jstor.org/stable/24407382

Eddy, Matthew Daniel, 2003, *The 'ingenious' rev. Dr. John Walker: chemistry, mineralogy and geology in enlightenment Edinburgh (1740–1800)*, Durham theses, Durham University. Available at Durham E-Theses Online: http://etheses.dur.ac.uk/4040/

Emerson, R., 1981, 'The Philosophical Society of Edinburgh 1748–1768', *British Journal for the History of Science*, 14(2), pp. 133–176, www.jstor.org/stable/4025944

Emerson, Roger L., 1985, 'The Philosophical Society of Edinburgh 1768–1783', *British Journal for the History of Science*, Vol. 18, No. 3 (Nov. 1985), pp. 255–303, https://www.jstor.org/stable/4026382

Emerson, Roger L., 1988, 'The Scottish Enlightenment and the End of the Philosophical Society of Edinburgh', *British Journal for the History of Science*, Vol. 21, No. 1 (Mar. 1988), pp. 33–66, https://www.jstor.org/stable/4026861

England, Philip C., Molnar, Peter, & Richter, Frank M., 2007, 'Kelvin, Perry and the Age of the Earth', *American Scientist*, Vol. 95, Issue 4 (Jul/Aug 2007), pp. 342–9, https://www.americanscientist.org/article/kelvin-perry-and-the-age-of-the-earth

Eyles, V., 1947, 'James Hutton (1726–97) and Sir Charles Lyell (1797–1875)', *Nature* 160, pp. 694–95, https://doi.org/10.1038/160694a0

Eyles, V.A. & Eyles, J.M., 1951, 'Some geological correspondence of James Hutton', *Annals of Science*, 7:4, pp. 316–39

Ferguson, Adam, 1801, 'Minutes of the Life and Character of Joseph Black, MD', *Transactions of the Royal Society of Edinburgh*, Vol. 5

Flinn, Derek, 'James Hutton and Robert Jameson', *Scottish Journal of Geology*, 16, pp. 251–58, 1 October 1980, https://doi.org/10.1144/sjg16040251

Gelfand, Toby, 2004, 'Walking the Paris hospitals: diary of an Edinburgh medical student, 1834–1835', *Medical History Supplement* 2004; (23): pp. vii–211

Gerstner, Patsy A., 1968, 'James Hutton's Theory of the Earth and His Theory of Matter', *Isis*, Vol. 59, No. 1 (Spring, 1968), pp. 26–31, https://www.jstor.org/stable/227849

Gerstner, Patsy A., 1971, 'The Reaction to James Hutton's Use of Heat as a Geological Agent', *British Journal for the History of Science*, Dec. 1971, Vol. 5, No. 4 (Dec. 1971), pp. 353–62, http://www.jstor.com/stable/4025379

Gohau, Gabriel, 1997, 'Hommage à James Hutton, à l'occasion du bicentenaire de sa mort', *Travaux du Comité français d'Histoire de la Géologie*, Comité français d'Histoire de la Géologie, 3ème série (tome 11), pp. 113–25, hal-00932547

Hair, P.E.H., 2000, 'Slavery and liberty: The case of the Scottish Colliers', *Slavery & Abolition*, 21:3, pp. 136–51, DOI: 10.1080/01440390008575324

Hall, Sir James, 1805, 'The Effects of Compression in Modifying the Action of Heat', *Transactions of the Royal Society of Edinburgh*, Vol. 6, (1812), p. 71

Ingenhousz, Jan, 1779, *Experiments upon Vegetables, Discovering Their great Power of purifying the Common Air in the Sun-shine, and of Injuring it in the Shade and at Night. To Which is Joined, A new Method of examining the accurate Degree of Salubrity of the Atmosphere*, London

Jacques, Jean, 1985, 'Le Cours de chimie de G.-F. Rouelle recueilli par Diderot', *Revue d'histoire des sciences*, Vol. 38, No. 1, pp. 43–53, https://doi.org/10.3406/rhs.1985.3992

Jones, Jean, 1982, 'James Hutton and the Forth and Clyde Canal', *Annals of Science*, 39:3, pp. 255–263

Jones, Jean, 1983, 'James Hutton: Exploration and Oceanography', *Annals of Science*, 40:1, pp. 81–94, DOI: 10.1080/00033798300200131

Jones, Jean, 1984, 'The geological collection of James Hutton', *Annals of Science*, 41:3, pp. 223–44, DOI: 10.1080/00033798400200241

Jones, Jean, 1985 'James Hutton's agricultural research and his life as a farmer', *Annals of Science*, 42:6, pp. 573–601

Jones, Jean, Torrens, Hugh S. & Robinson, Eric, 1994, 'The correspondence between James Hutton (1726–1797) and James Watt (1736–1819) with two letters from Hutton to George Clerk-Maxwell (1715–1784): Part I', *Annals of Science*, 51.6, pp. 637–53

Jones, Jean, Torrens, Hugh S. & Robinson, Eric, 1995, 'The correspondence between James Hutton (1726–1797) and James Watt (1736–1819) with two letters from Hutton to George Clerk-Maxwell (1715–1784): Part II', *Annals of Science*, 52, pp. 357–82

Kirwan, Richard, 1793, 'Examination of the Supposed Igneous Origin of Stony Substances', *Transactions of the Royal Irish Academy*, Vol. 5 (1793 /1794), pp. 51–82

Kirwan, Richard, 1797, 'On the Primitive State of the Globe and Its Subsequent Catastrophe', *Transactions of the Royal Irish Academy*, Vol. 6 (1797), pp. 233–308

Lehman, Christine, 2004, *Le cours de chimie de Guillaume-François Rouelle*, http://rhe.ish-lyon.cnrs.fr/cours_magistral/expose_rouelle/expose_rouelle_complet.php

Lehman, Christine, 2011, 'Les multiples facettes des cours de chimie en France au milieu du XVIIIe siècle', *Histoire de l'éducation*, 130, pp. 31–56

Levere, Trevor H. & Turner, Gerard L'E, 2002, *Discussing Chemistry and Steam – The Minutes of a Coffee House Philosophical Society 1780–1787* – Oxford Scholarship Online, DOI:10.1093/acprof:oso/ 9780198515302.001.0001

McElroy, D.D., 1952, *The Literary Clubs and Societies of 18th century Scotland*, dissertation, University of Edinburgh

McIntyre, Donald B., 1997, 'James Hutton's Edinburgh; the historical social and political background', *Earth Sciences History*, Vol. 16, No. 2, pp. 100–157

McIntyre, Donald B., 1999, 'James Hutton's Edinburgh: a précis', Geological Society, London, Special Publications, Vol. 150, pp. 1–12, https://doi.org/10.1144/GSL.SP.1999.150.01.01

Morrell, J. B., 1971, 'Professors Robison and Playfair, and the 'Theophobia Gallica': Natural Philosophy, Religion and Politics in Edinburgh, 1789–1815', *Notes and Records of the Royal Society of London*, Vol. 26, No. 1, pp. 43–63. www.jstor.org/stable/531052

Multhauf, Robert P., 1965, 'Sal Ammoniac: A Case History in Industrialization', *Technology and Culture*, Vol. 6, No. 4 (Autumn, 1965), pp. 569–86, Johns Hopkins University Press and the Society for the History of Technology

Munro, Rolland, papers and notebooks, University of Edinburgh Centre for Research Collections, Coll-804

Norwick, Stephen, 2002, 'Metaphors of nature in James Hutton's *Theory of the earth with proofs and illustrations*', *Earth Sciences History*, Vol. 21, No. 1, pp. 26–45, www.jstor.org/stable/24137170

O'Rourke, J.E., 1978, 'A Comparison of James Hutton's Principles of Knowledge and Theory of the Earth', *Isis*, Vol. 69, No. 1 (Mar, 1978), pp. 4–20, https://www.jstor.org/stable/230605

Pearson, Paul N., Supplementary Information to accompany the 'In Retrospect' article on James Hutton's *The Principles of Knowledge*, http://www.blc.arizona.edu/courses/schaffer/449/In%20Retrospect%20Supplemental.pdf

Perrin, Carleton, 1983, 'Joseph Black and the absolute levity of phlogiston', *Annals of Science*, 40:2, pp. 109–37, DOI: 10.1080/00033798300200151

Perry, John, 1895, 'On the Age of the Earth', *Nature* 51, pp. 341–42, https://doi.org/10.1038/051341b0

Playfair, J., 1801, 'Minutes of the life and character of Dr. Joseph Black', *Transactions of the Royal Society of Edinburgh*, Vol. 5

Porter, Roy S., 1978, 'Philosophy and Politics of a Geologist: G. H. Toulmin (1754–1817)', *Journal of the History of Ideas*, Vol. 39, No. 3, 1978, pp. 435–50, www.jstor.org/stable/2709387

Porter, Roy, 1978, 'George Hoggart Toulmin's theory of man and the earth in the light of the development of British geology', *Annals of Science*, 35:4, pp. 339–52, DOI: 10.1080/00033797800200281

Principe, Lawrence M., 2014, 'The End of Alchemy? The Repudiation and Persistence of Chrysopoeia at the Académie Royale Des Sciences in the Eighteenth Century', *Osiris*, Vol. 29, No. 1, 2014, pp. 96–116, www.jstor.org/stable/10.1086/678099. Accessed 8 Sept. 2021

Raper, Horace W., 1979, 'Andrew Balfour 1737–1782', *Dictionary of North Carolina Biography*, University of North Carolina Press

Rappaport, Rhoda, 'G-F. Rouelle: An Eighteenth-Century Chemist and Teacher', *Chymia*, Vol. 6 (1960), pp. 68–101, University of California Press

Roberts, Lissa & van Driel, Joppe, 'The Case of Coal', in Roberts, Lissa and Werrett, Simon, 2017, *Compound Histories: Materials, Governance and Production*, 1760–1840, Brill

Rognstad, Matthew, *Lord Kelvin's Heat Loss Model as a Failed Scientific Clock*, http://apps.usd.edu/esci/creation/age/content/failed_scientific_clocks/kelvin_cooling.html

Ross, Kevin, 2011, *James Hutton's Metaphysics, Theory of Language and Science, in the Scottish Enlightenment*, PhD thesis, University of Edinburgh

Rouelle, G-F., *Leçons de Chimie de M. Rouelle, Notes prises au cours en 1754 et 1755, corrigées en 1757 et 1758*, D'après le manuscrit, Fr. N° 12303–12304, conservé à la Bibliothèque nationale de France, TOME 2

Shapin, Steven, 1971, *The Royal Society of Edinburgh: A Study of the Social Context of Hanoverian science*, University of Pennsylvania, PhD thesis

Tann, Jennifer, 2004, *Watt, James (1736–1819)*, Oxford DNB, https://doi.org/10.1093/ref:odnb/28880

Thackray, John, 1998, 'Charles Lyell and the Geological Society', Geological Society, London, Special Publications, 143, pp. 17–20, https://doi.org/10.1144/GSL.SP.1998.143.01.03

Thomson, George, 1950, 'James Watt and the Monkland Canal', *Scottish Historical Review*, Vol. 29, No. 108, Part 2, pp. 121–133

Thomson, William, 1868, 'On Geological Time', Address delivered before the Geological Society of Glasgow, February 27, 1868. Popular Lectures and Addresses, Vol. ii, p. 10

Tomkeieff, S., 1947, *James Hutton and the philosophy of geology*, http://trned.lyellcollection.org/ at University of Edinburgh

Tomkeieff, S., 1953, 'Hutton's Unconformity, Isle of Arran', *Geological Magazine*, 90(6), pp. 404–08, DOI:10.1017/S0016756800065924

Tomkeieff, S., 1962, 'Unconformity – An Historical Study', *Proceedings of the Geologists' Association*, Vol. 73, Issue 4, pp. 383–417, ISSN 0016-7878, https://doi.org/10.1016/S0016-7878(62)80031-5

Torrens, Hugh, 2008, 'Geological Pioneers in the Marches', *Proceedings of the Shropshire Geological Society*, 13, pp. 65–76

Wisniak, Jaime, 2003, 'Guillaume François Rouelle', *Educación Química*, January 2003, DOI: 10.22201/fq.18708404e.2003.4.66232

Withers, Charles W.J., 1989: 'William Cullen's Agricultural Lectures and Writings and the Development of Agricultural Science in Eighteenth-Century Scotland', *Agricultural History Review*, Vol. 37, No. 2, pp 144–56

BOOKS AND BOOK CHAPTERS

Bakewell, Robert, 1833 *An introduction to geology*, Longman

Baxter, Stephen, 2003, *Revolutions in the Earth: James Hutton and the True Age of the World*, W&N

Bower, Alexander, 1817, *History of the University of Edinburgh*, Vol. II, Oliphant, Waugh and Innes

Broadie, Alexander, 2018, *The Scottish Enlightenment*, Birlinn

Buffon, Georges Louis Leclerc, Comte de, 1781, *Natural history: general and particular, by the Count de Buffon, translated into English*, Creech 1781 https://quod.lib.umich.edu/e/ecco/004880992.0001.001/1:2?rgn =div1;view=fulltext;q1=Buffon

Carlyle, Alexander, 1860, *Autobiography of the Rev. Dr. Alexander Carlyle, Minister of Inveresk, containing memorials of the men and events of his time*, W. Blackwood

Chambers, Robert, 1868, *Traditions of Edinburgh*, Chambers

Clerk, Sir John, 1892, *Memoirs of the Life of Sir John Clerk of Penicuik: Baronet, Baron of the Exchequer, Extracted by Himself from His Own Journals, 1676–1755*, Constable for the Scottish History Society

Cockburn, Henry, 1977, *Memorials of his time*, James Thin

Cozens-Hardy, Basil, 1950, *The Diary of Silas Neville 1767–1788*, Oxford University Press

Craig, G.Y. (ed.), McIntyre D.B., Waterston C.D., 1978, *James Hutton's Theory of the Earth: The Lost Drawings*, Scottish Academic Press, in Association with the Royal Society of Edinburgh and the Geological Society of London

Crichton-Browne, Sir James, 1926, *Victorian Jottings from an Old Commonplace Book*, Etchells & Macdonald

Cunningham, Andrew, Grell, Ole Peter, & Arrizabalaga, Jon (eds), 2016, *Centres of Medical Excellence? Medical Travel and Education in Europe, 1500–1789*, Routledge

De Luc, Jean André, 1791, *Letters on the Physical History of the Earth, Addressed to Professor Blumenbach: Containing Geological and Historical Proofs of the Divine Mission of Moses*, C.J.G. & F. Rivington, 1831

Dean, Dennis R., 1992, *James Hutton and the History of Geology*, Cornell University Press

Dean, Dennis R., 1997, *James Hutton in the field and in the study*, Delmar, NY

Dowds, T.J., 2003, *The Forth & Clyde Canal, a history,* Tuckwell Press

Drescher, Horst W. (ed.), 1999, *Literature and literati: the literary correspondence and notebooks of Henry Mackenzie,* Vol. 2, Notebooks 1763–1824, Peter Lang

Faujas-de-St.-Fond, Barthélemy, 1799, *Travels in England, Scotland, and the Hebrides,* James Ridgeway

Furniss, Tom, 2018, 'Astonishing Productions of Volcanic Combustion: Barthélemy Faujas de Saint-Fond's Travels in England, Scotland, and the Hebrides (1784, 1799)', in *Discovering the Footsteps of Time,* Edinburgh University Press

Geikie, Archibald, 1858, *The Story of a Boulder: or gleanings from a notebook of a field geologist,* Thomas Constable

Geikie, Archibald, 1895, *Memoir of Sir Andrew Crombie Ramsay,* Macmillan, London

Geikie, Archibald, 1899, *Theory of the Earth: With Proofs and Illustrations by James Hutton,* Geological Society

Geikie, Archibald, 1905, *The founders of geology,* Macmillan, London

Geikie, Archibald, 1924, *A long life's work, an autobiography,* Macmillan, London

Gelfand, Toby, 2016, 'Paris 'certainly the best Place for learning the practical part of Anatomy and Surgery' in Cunningham, Grell & Arrizabalaga (eds), *Centres of Medical Excellence? Medical travel and education in Europe 1500–1789,* Routledge

Gilhooley, James, 1988, *A directory of Edinburgh in 1752,* Edinburgh University Press

Gould, Stephen Jay, 1987, *Time's Arrow, Time's Cycle: Myth and Metaphor in the Discovery of Geological Time,* Harvard University Press

Heron, Alexander, 1903, *The Merchant Company of Edinburgh 1681–1902,* T & T Clark

Howard, Philip, 1797, *The scriptural history of the earth and of mankind,* http://hdl.handle.net/10427/009192

Jones, Peter, 1984, 'An Outline of the Philosophy of James Hutton', in Hope, V. (ed.) *Philosophers of the Scottish Enlightenment,* Edinburgh University Press

Kay, John, 1837, *A series of original portraits and caricature etchings with biographical sketches and illustrative anecdotes,* Hugh Paton, Carver and Gilder

Kay, John, 1842, *A Series of original portraits and caricature etchings,* A & C Black

Kerr, Robert, 1992, *Memoirs of the life, writings, and correspondence of William Smellie*, Vol. II, Thoemmes Press

King-Hele, Desmond, 2007, *The collected letters of Erasmus Darwin*, Cambridge University Press

Kölbl-Ebert, Martina, 2009, *Geology and Religion: A History of Harmony and Hostility*, Geological Society of London

Leddra, Michael, 2010, *Time Matters: Geology's Legacy to Scientific Thought*, Blackwell

Lyell, Charles, 1830–33, *Principles of Geology*, John Murray, https://library. si.edu/digital-library/book/principlesgeolovol1lyel

Marshall, Rosalind K., 2015, *The Edinburgh Merchant Company*, John Donald

McIntyre, Donald B., 1963, 'James Hutton and the philosophy of geology', in *The Fabric of Geology*, Addison Wesley

Merolle, Vincenzo, 1995, *The Correspondence of Adam Ferguson*, 2 Vols, William Pickering

Mitchell, James, 1823, *A Dictionary of Chemistry, Mineralogy, and Geology*, Phillips

Muirhead, J.P., & Watt, J. 1854, *The Origin and Progress of the Mechanical Inventions of James Watt*, John Murray

Murray, John, 1802, *A Comparative View of the Huttonian and Neptunian Systems of Geology: In Answer to the Illustrations of the Huttonian Theory of the Earth, by Professor Playfair*, Ross and Blackwood

O'Brian, Patrick, 1987, *Joseph Banks, a life*, Collins Harvill

Newton, Gill, 2012, *Marriage among Londoners before Hardwicke's Act of 1753: when, where and why?* Cambridge Group for the History of Population and Social Structure

Nicolas, Pierre-François, 1787, *Précis des leçons publiques de chimie et d'histoire naturelle*, Henry Haener

Playfair, J., 1803, *Biographical Account of James Hutton, M.D. F.R.S. Ed* (Cambridge Library Collection – Earth Science, pp. 1–61), Cambridge University Press, DOI:10.1017/CBO9780511973253.001

Playfair, John, 1802 *Illustrations to the Huttonian theory of the earth*, Cadell and Davies, William Creech, https://www.biodiversitylibrary.org/item/107595#page/7/mode/1up

Porter, Roy, 1977, *The Making of Geology: Earth Science in Britain, 1660–1815*, Cambridge University Press

Rae, John, 1895, *Life of Adam Smith*, Macmillan, https://oll.libertyfund.org/titles/1411#Rae

Ramsay, William, 1918, *The life and letters of Joseph Black, MD*, Constable

Rasmussen, Dennis C., 2017, *The Infidel and the Professor: David Hume, Adam Smith, and the Friendship that Shaped Modern Thought*, Princeton University Press

Repcheck, Jack, 2003, *The man who found time: James Hutton and the discovery of the ancient earth*, Simon & Schuster

Robinson, Eric & McKie, Douglas, eds, 1970. *Partners in science: letters of James Watt and Joseph Black*, London: Constable

Rosner, Lisa M., 1991, *Medical education in the Age of Improvement: Edinburgh students and apprentices, 1760–1826*, Edinburgh University Press

Ross, William, 1947, *The Royal High School*, Oliver & Boyd

Rudwick, Martin J.S., 2007, *Bursting the Limits of Time: The Reconstruction of Geohistory in the Age of Revolution*, University of Chicago Press

Smellie, William, 1782, *Account of the Institution and Progress of the Society of the Antiquaries of Scotland*, William Creech; and Thomas Cadell

Smiles, Samuel, 1865, *The lives of Bouton and Watt*, John Murray, https://www.gutenberg.org/files/52069/52069-h/52069-h.htm

Smith, J., 2011, *Memoir and Correspondence of the Late Sir James Edward Smith, M.D.* (Cambridge Library Collection – Botany and Horticulture) (P. Smith, ed.). Cambridge: Cambridge University Press, DOI:10.1017/CBO9781139095488

Smitten, Jeffrey R., 2016, *Life of William Robertson: Minister, Historian, and Principal*, Edinburgh University Press

Smout, T.C., 1983, 'Where had the Scottish economy got to by 1776?', in Hunt, I & Ignatieff, M (eds), *Wealth and Virtue*, Cambridge University Press

Trotter, James, 1911, *The Royal High School*, Pitman

Walker, John, 1812, *Essays on Natural History and Rural Economy*, Longman, Hurst, Rees, and Orme

Wallis, P.J., Wallis, R.V., & Whittet, TD, 1985, *Eighteenth century medics: subscriptions, licences, apprenticeships*, Project for Historical Bibliography, Newcastle Upon Tyne

Watson, Thomas, 1894, *Kirkintilloch, town and parish*, John Smith, Glasgow

Whitehurst, John, 1786, *An Inquiry into the Original State and Formation of the Earth*, 2nd edn, Bent, London

Williams, John, 1789, *The natural history of the mineral kingdom*, Vol. 1, printed in Edinburgh for the author

Wilson, Leonard G., 1998, *Lyell, the man and his times*, Geological Society, London, Special Publications, 143, pp. 21–37, 1 January 1998, https:// doi.org/10.1144/GSL.SP.1998.143.01.04

Wollstonecraft, Mary, 1792, *A Vindication of the Rights of Woman with Strictures on Political and Moral Subjects*, J. Johnson, https://oll.liberty-fund.org/titles/126

Wood, Marguerite & Marwick, James D,1869, *Extracts From the Records of the Burgh of Edinburgh*, Scottish Burgh Records Society

Notes and references

CHAPTER 1

1. Black to Watt, 13 June 1797, Robinson and McKie, pp. 276–7
2. Isabella Hutton's name is variously spelt Isobel, Isabell and Isabella in documents of the time. For consistency I use the spelling given by Playfair.
3. Stevenson, R.L., *Memories, Portraits, Essays and Records*, e-book, Jazzybee Verlag, 2014, p. 397
4. Playfair 1803
5. Black to Watt, 1 October 1772, Robinson and McKie 1970, p. 33
6. Ferguson 1801, p. 114
7. Playfair 1803, p. 59
8. Hutton to George Clark, July 1755, NRS GD18/5749
9. Playfair 1803, p. 6
10. *Ibid.*, p61
11. See, for example, Cozens-Hardy 1950, pp. 137–43
12. Munro papers
13. *Scots Magazine*, 26 July 1780
14. *Caledonian Mercury*, 5 December 1789
15. Jean Jones 1985, p. 584
16. Repcheck 2003, p. 88
17. Burns wrote a poem about it, 'The Fornicator' https://www.bbc.co.uk/arts/robertburns/works/the_fornicator/
18. See, for example, NRS CH2/122/12, pp. 4, 129, 147, or CH2/718/59, CH2/122/75
19. NRS CH2/127, CH2/128
20. Torrens papers, p. 358, handwritten note probably by Jean Jones, or Joan Eyles, History of Geology Archives, University of Oklahoma

Libraries. QED is an abbreviation of Quod Erat Demonstrandum – it was demonstrated.

21. Black to Watt, 13 June 1797, *op. cit.*
22. I am grateful to Ann Farish of Flimby for this information.
23. Playfair read the first version of his biographical account of Hutton's life to the Royal Society of Edinburgh in 1803, six years after Hutton's death.
24. Jean Jones 1985, p. 584

CHAPTER 2

1. J.J. Brown, quoted in Dingwall 1989, vol. 1, p. 343
2. Wood & Marwick 1869, pp. 24, 203
3. NRS GD436/3/36
4. NRS Hearth tax records for Midlothian, vol. 2 (Edinburgh city) E69 /16/2/99
5. Dingwall 1989, vol. 2, appendix 3
6. Gosford Close was demolished in the 1830s to make way for George IV Bridge.
7. Roll Book of the Merchant Company of Edinburgh
8. Heron 1903, pp. 65–6
9. Marshall 2015, p. 11
10. *Ibid.*, p. 15
11. NRS CH2/127, 9 September 1723
12. TCM SL1/1/48 p. 248; SL1/1/49 p. 73, p. 369; SL1/1/50; p. 353; SL1 /1/52 p. 26
13. NRS CC8/8/92
14. Ross, W., 1947 p. 25
15. TCM 14 February 1739
16. Ross, W., 1947, p. 47
17. Cockburn 1977, p. 13
18. Ross, W., 1947, Trotter 1911
19. Cockburn 1977, p. 4
20. University of Edinburgh Matriculation Album Vol. II, 1704–62 EUA IN1/ADS/STA/2/2
21. Morrell 1971, p. 159
22. Carlyle 1860, p. 31
23. Bower 1817, pp. 247–81

24. Brackett 1935, p. 41
25. Playfair 1803, p. 2
26. EUA IN1/ADS/STA/1/1

CHAPTER 3

1. See Maclaurin's entry in the Oxford Dictionary of National Biography
2. See O'Rourke 1978, and Ross, K., 2011
3. NRS GD436/3/36. By 1732 Lady Cringletie was using Margaret Hamilton as her tailor.
4. NA IR1/50 p. 186, 10 August 1743
5. Playfair 1803, p. 3
6. Donovan 1978, p. 6
7. EU CRC Monro's list. Plummer's list has not survived, so we don't know if Hutton also attended his chemistry classes. He did not formally matriculate.
8. Knoeff, R., 'Hermann Boerhaave at Leiden', in Cunningham et al. 2016
9. Now the site of the Deacon Brodie's Tavern.
10. McIntyre 1963, p. 352
11. Wallis, Wallis & Whittet 1985, p. 585, 1261 give the date as 27 September, but Dr Young's duty payment was recorded on 14 May, with the apprenticeship recorded as beginning 4 April. NA IR1/50 p. 297
12. https://library.rcsed.ac.uk/about-us/blog/archive/surgeons-apprentices-bodysnatching-and-other-immoral-behaviour-in-18th-century-edinburgh
13. McIntyre 1963, p. 354
14. Rosner 1991, p. 40
15. Cozens-Hardy 1950, p. 156
16. *Ibid.*, p. 144
17. EU CRC Monro's list
18. NRS GD18/2331, 16 April 1745
19. Oxford DNB https://doi.org/10.1093/ref:odnb/5617; Brown 2008, p. 15
20. NRS GD18/5749
21. Clerk 1892, p. 188

CHAPTER 4

1. EU CRC Index to the Register of Edinburgh apprentices 1701–1755
2. Playfair 1803, p. 3
3. Gelfand 2016
4. https://fr.wikipedia.org/wiki/Jardin_royal_des_plantes_
   médicinales#Le_règne_de_Buffon_1738–88
5. Gelfand 2004
6. Gelfand 2016, p. 229
7. McIntyre 1963, p. 354
8. Rappaport 1960
9. Lehman 2004
10. Lehman 2011
11. Leçons de Chimie de M. Rouelle, Fr. N° 12303–12304, Bibliothèque
    Nationale de France, Tome 2, http://www.annales.org/archives/
    cofrhigeo/RouelleCoursChimie.pdf
12. Principe 2014, p. 114
13. Wisniak 2003, p. 243
14. Geikie 1905, p. 142
15. Boot's name is mentioned on Hutton's dissertation, but by the time he
    was awarded his degree he appears to have been lodging with the
    Widow Sass on the Langebrug, a covered canal about five minutes'
    walk from the university.
16. Archieven van Senaat en Faculteiten, 1727–55, ASF 14, p. 380,
    University of Leiden
17. Little is known about Stevenson and his relation to Hutton. He gradu-
    ated in medicine in 1710 at the University of Harderwijk (where
    Boerhaave had earlier been an undergraduate) and apparently also
    studied at the University of Glasgow. He became a Licentiate of the
    Royal College of Physicians of Edinburgh in 1729 and was elected a
    Fellow later the same year.
18. Professor of the Institutes of Medicine 1726–47
19. Donovan & Prentiss 1980, pp. 24–27. This paper also includes
    Hutton's original thesis in Latin and a translation into English.
20. *Ibid.,* p. 43
21. *Ibid.,* p. 8
22. Norwick 2002, p. 34
23. Broadie 2018, p. 201

24. http://www.newtonproject.ox.ac.uk/view/contexts/CNTX00001
25. Donovan & Prentiss 1980, p. 31
26. *Ibid.*, p. 27

CHAPTER 5

1. Playfair 1803, p. 4
2. *Ibid.*
3. Church of England records do not show a James Hutton born or baptised in London in the period 1747–50, nor do the surviving non-conformist church records, nor those of the Scots Church at Founder's Hall, Lothbury 1716–73. Tax records show a James Hutton living at Farringdon, Fetter Lane and Wapping during dates in 1749, but there is no evidence that this is the Edinburgh James Hutton.
4. A James Hutton was admitted to Christ's Hospital school for boys in 1763, but he was aged 8 and his father was William Hutton of Middlesex. There is no record of a James Hutton in the Raines Foundation School.
5. Royal Sun Alliance records held at the London Metropolitan Archives show James Hutton as insuring properties from 1782–9.
6. *Ibid.*
7. JH to George Clark Esq July (?) 1755, NRS GD18/5749
8. Marriage registers retrieved via Ancestry.co.uk
9. NRS CH2/122/12/4, p. 181
10. Retrieved from Scotlandspeople.gov.uk
11. London Metropolitan Archives, see also Newton 2012
12. Ferguson 1801, p. 114
13. Playfair wrongly calls him 'James' Davie
14. NRS RS3/98
15. NRS RS19/14
16. Withers 1989, p. 144
17. *Ibid.*, p. 148
18. Hutton, *Elements*, p. 2
19. *Ibid.*, p. 6
20. NRS GD206/2/268
21. Hutton, *Elements*, p. 11
22. Hutton spells the name Dibol, although Playfair spells it as Dybold.

23. Playfair 1803, p. 5
24. Quoted by Playfair, but now lost
25. Eyles & Eyles 1951, pp. 323–4
26. *Ibid.*
27. Hutton 1795, *Theory of the Earth*, p. 167

CHAPTER 6

1. https://www.nrscotland.gov.uk/files//research/census-records/
   websters-census-of-1755-scottish-population-statistics.pdf
2. NRS GD18/5749 July 1755
3. Hutton, *Elements*, p. 471
4. Jones 1985, p. 586
5. NRS GD18/5749 July 1755
6. *Ibid.* I have altered the punctuation to make it more readable.
7. Playfair 1803, p. 6
8. NRS GD206/2/554/21
9. NRS GD206/2/269, pp. 1–14
10. Her uncle John Pringle was also a doctor, but he was in London at the
    time and surely she would have called him by his name?
11. NRS GD18/5749
12. Gilhooley 1988, p. 27
13. NRS CS271/6948, Hutton v Livingstone 1766
14. Hutton, *Elements*, p. 450
15. *Ibid.*, p. 8
16. Jones 1985, p. 587
17. Hutton, *Elements, passim*
18. Jones 1985, p. 590
19. Hutton, *Elements*, p. 371
20. Hutton, *Elements*, pp. 354. He explained it by reference to the phlogis-
    ton theory
21. *Ibid.*, p. 774
22. Jones 1985, p. 588
23. Hutton, *Elements*, p. 869
24. *Ibid*, pp. 825–31
25. NRS GD1/1432/1/26
26. The couple adopted the surname Clerk Maxwell and were the great-
    grandparents of the physicist James Clerk Maxwell.

27. Eyles & Eyles 1951, p. 326
28. Letters to Sir John Strange, quoted in Eyles & Eyles 1951
29. Hutton, *Elements*, p. 9
30. *Ibid.*, p. 331

CHAPTER 7

1. NA IR1/50 p. 346 1 June 1743
2. NRS CS271/60776
3. *Ibid.*, 15 October 1762
4. https://maps.nls.uk/view/74400010
5. NRS SIG1/48/22
6. Playfair 1803, p. 8
7. See https://www.scottisharchivesforschools.org/naturalScotland/Cudbear. asp and https://www.theglasgowstory.com/story/?id=TGSBE
8. Hutton, *Elements*, p. 331
9. Playfair dates the partnership at 1765, but says that Davie had run the business on his own for the previous nine years. This seems unlikely. Hutton was not in Edinburgh and the manufacture of the chemical on an industrial scale would have needed a substantial factory. The evidence from note 1, above, is that nothing was built by Davie prior to 1762. It also does not explain why Davie felt it necessary to include Hutton in the partnership if it was already an established business.
10. Rouelle lectures BNP 12303–4. The original text is: 'J'ai trouvé le moyen d'en faire de toutes pièces en combinant l'acide vitriolique avec une terre végétale de quelque bois fossile.'
11. Multhauf 1965
12. Nicolas 1787, pp. 125–7
13. Cozens-Hardy 1950, p. 153
14. Jacques 1985, p. 48
15. Multhauf 1965, p. 572
16. *Ibid.*
17. Clow & Clow 1947, p. 425
18. Clow & Clow 1945, p. 45
19. The 1794 street directory gives the address as 171 Nicolson Street. Modern numbering puts that address on the west side of the street, but the factory appears to have been on the east side.
20. Mitchell 1823, p. 163

21. Joseph Black to Andrew Stuart, Edinburgh, 25 January 1783, quoted in Clows & Clows, 1947, p. 426
22. https://www.oxforddnb.com/view/10.1093/ref:odnb/9780198614128.001.0001/odnb-9780198614128-e-5750
23. https://digital.nls.uk/directories/browse/archive/83158635

Chapter 8

1. Smout 1983, p. 58
2. Emerson 1985
3. Playfair 1803, p. 57
4. Ramsay 1918, p. 115
5. *Ibid.*, p. 117
6. *Ibid.*, p. 83
7. https://en.wikipedia.org/wiki/James_Tytler and https://en.wikipedia.org/wiki/Vincenzo_Lunardi
8. Playfair 1803, p. 12
9. Dowds 2003, pp. 24–31
10. NRS BR/FCN/1 21 Dec 1767
11. Dowds 2003, p. 33
12. Watson, p. 87
13. *Ibid.*
14. NRS BR/FCN/1 28 June 1768; Jones 1982, pp. 259–60
15. NRS BR/FCN/1 4 June 1768
16. NRS BR/FCN/ 22 April 1769
17. NRS BR/FCN/ 22 Sept 1769

Chapter 9

1. Hutton to Sir John Strange, quoted in Eyles & Eyles 1951, p. 326
2. https://canmore.org.uk/site/52234/edinburgh-canongate-st-johns-land, https://www.ed.ac.uk/education/about-us/maps-estates-history/estates/st-johns, https://www.stjohnscotland.org.uk/the-order/the-order-in-scotland
3. Crichton-Browne 1926, pp. 1–3
4. Cozens-Hardy 1950, p. 145
5. *Ibid.*
6. *Ibid.*, p. 153

7. Emerson 1985, p. 255
8. Playfair 1803, p. 12
9. *Transactions of the Royal Society of Edinburgh*, Vol. 2, 1790
10. Hutton 1777
11. Anon., 1777. *Remarks on considerations on the nature, quality and distinctions, of coal and culm, &c*, London.
12. Roberts & van Driel 2017, pp. 63–5
13. Hutton to Sir John Strange, quoted in Eyles & Eyles 1951, pp. 317–18
14. Eyles and Eyles believe that Hutton may have been the first in Britain to have drawn a geological map, *ibid.*, p. 334
15. *Ibid.*, pp. 322–3
16. *Ibid.*
17. Hutton to Lind, early 1772, in Jones 1983, p. 83
18. O'Brian 1987, see Ch. 6
19. Hutton to Banks, undated in Jones 1983, p. 85
20. I am grateful to Professor Alan Werritty for this information from his analysis of Hutton's *Elements of Agriculture*

Chapter 10

1. Watt to Margaret Watt, 5 April 1767, Boulton and Watt MSS, quoted in Tann 2004
2. Watt to Dr Small, November 1772, quoted in Thomson 1950
3. Tann 2004, *op. cit.*
4. Quoted in Jones, Torrens & Robinson 1994, p. 641
5. Jones, Torrens & Robinson, 1994, p. 639
6. *Transactions of the Royal Society*, 13 December 1787, https://royalsoci-etypublishing.org/toc/rstl/1788/78
7. Letter Erasmus Darwin to Hutton, 3 July 1778, King-Hele 2007, pp. 78–9
8. Hutton to George Clerk, July 1774, quoted in Jones, Torrens & Robinson 1994, p. 642
9. Hutton to George Clerk, August 1774, quoted in Jones, Torrens & Robinson 1994, p. 648
10. See, e.g., Smith 2011, p. 44, and Cozens-Hardy 1950 *passim*.
11. Hutton to George Clerk, August 1774, quoted in Jones, Torrens & Robinson 1994, p. 648

12. Jones, Torrens & Robinson 1994, footnotes p. 643. The full note reads 'A slice of cucumber' appears to be the equivalent of the more usual expression 'a slice of a cut loaf', meaning fornication with a married woman. 'A' instead of 'a', the Scots form of 'all'. 'C' for 'cunt'? 'Provocative' = Aphrodisiac.
13. Hutton to James Watt, undated but endorsed by Watt 'Dr Hutton, Bath, 1774'. Quoted in Jones, Torrens & Robinson 1994, p. 646
14. Hutton to James Watt, October 1774, quoted in Jones, Torrens & Robinson 1995, p. 359
15. Torrens 2008, p. 68
16. Hutton to James Watt, undated but endorsed by Watt 'Dr Hutton, 1774.' Postmarked Manchester, quoted in Jones, Torrens & Robinson 1994, p. 651
17. Two of the hazards faced by Odysseus in Homer's epic poem. Hutton may have read it in Greek, but by the time he was at school there were at least a dozen English translations.
18. Hutton to James Watt, undated but endorsed by Watt 'Dr Hutton, 1774', quoted in Jones, Torrens & Robinson 1995, Part II, p. 359
19. A reference to the satire 'A Modest Proposal,' of Jonathan Swift, 1729
20. Hutton to James Watt, October 1774, quoted in Jones, Torrens & Robinson 1995, p. 360
21. Smiles 1865, ch. X
22. Hutton to James Watt, December 1774, quoted in Jones, Torrens & Robinson 1995, p. 362
23. *Ibid.*
24. Smiles 1865, ch. X

CHAPTER 11

1. British Postal Service Appointment Books, 1737–1969, 8 May 1770 (retrieved from Ancestry.co.uk)
2. London, England, Church of England Marriages and Banns, 1754–1932, 5 August 1775 (retrieved from Ancestry.co.uk)
3. England & Wales, Prerogative Court of Canterbury Wills, 1384–1858 for William Smeeton, PROB 11: Will Registers 1777–80 Piece 1027: Collier, Quire Numbers 1–48 (1777) (retrieved from Ancestry.co.uk)

4. 'Consols', a common form of government debt were yielding 4.5% in 1778, up from 3.8% the previous year. *British Historical Facts*, 1760– 1830, Cook & Stevenson, Springer, 2016

5. James Hutton letter to Dr J. Hutton, 1778, BL Add MS 36297

6. Raper 1979. It is likely that both brothers owned slaves on their plantation https://www.ucl.ac.uk/lbs/person/view/2146653369

7. See Raper 1979. George Washington appointed her as postmistress, the first woman to hold the post.

8. Munro papers

9. https://royalsocietypublishing.org/doi/pdf/10.1098/rspb.1924.0036

10. University of Glasgow Archive: MS Gen 1035/151

11. Rae 1895, 0238 685

12. Drescher 1999, p. 215

13. Rae 1895, 0238 687

14. Chambers 1868, p. 338

15. McElroy 1952, p. 139

16. McElroy 1952, pp. 521–6

17. Rae 1895, 0238 679

18. *Transactions of the Royal Society of Edinburgh*, Vol. 5, read 10 January 1803

19. Playfair 1803, p. 60

20. Kay 1837, p. 49

21. Letter from Joseph Black to Dashkova (Edinburgh, 27 August 1787), EU CRC Ms.Gen.873/III, fols 36–8

CHAPTER 12

1. McElroy 1952, p. 335

2. Emerson 1988, pp. 36–7

3. Shapin 1971, p. 138

4. Smellie 1782, p. 4

5. Brown 2008

6. Shapin 1971, pp. 146, 157

7. *Ibid.*, p. 185

8. NLS Acc10000/1 RSE minutes 23 June 1783 – 6 July 1791

9. *Transactions of the Royal Society of Edinburgh*, Vol. I, 1788

10. It was not published until 1790 in the *Transactions of the Royal Society of Edinburgh*, Vol. II

11. Playfair 1803, p. 55

12. The Scottish National Portrait Gallery dates this picture to 1776, but since it is assumed to show the manuscript of the *Theory of the Earth*, it was more likely painted in the following decade. Professor David Mackie dates it c.1790. https://www.nationalgalleries.org/art-and-artists/2808/james-hutton-1726-1797-geologist, McIntyre 1999, p. 11

13. Mémoires d'agriculture, d'économie rurale et domestique/publiés par la Société royale d'agriculture de Paris, https://gallica.bnf.fr/Académie d'Agriculture de France

14. Quoted in Geikie 1905, p. 169

15. *Transactions of the Royal Society of Edinburgh*, Vol. 1, 1788, p. 41

16. Playfair 1803, footnote p. 28. Playfair calculated the difference as 1°F for each 250 feet of altitude.

17. Levere & Turner 2002, p. 171

18. Read 3 December 1787, published in *Transactions of the Royal Society of Edinburgh*, Vol. 2, 1790

19. Playfair 1803, p. 27

20. See Brewer, A.W., 1952, *Why does it rain?* Weather, Vol. 7, pp. 195–8, https://doi.org/10.1002/j.1477-8696.1952.tb01494.x

Chapter 13

1. For a description of Hutton's mineral collection and what happened to it after his death see Jones, 1984

2. Playfair 1803, pp. 12–13

3. Faujas de Saint-Fond 1799, pp. 229–30

4. Jones 1984, p. 223

5. Faujas de Saint-Fond 1799, p. 253

6. Dean 1992, p. 14

7. Furniss 2018, p. 151

8. See Leddra 2010

9. See Dean 1981, p. 188

10. Smellie was an important and influential publisher but led a colourful private life and was constantly short of money. He several times wrote to Hutton, ostensibly about health matters, but usually ending with an appeal for money. On 20 June 1793, for example: 'Now for the Devil, who is like to stick both in my throat and my pen; an hundred pounds

would greatly aid the operation of exercise and bathing, by removing some loads that lie heavy on my stomach.' Kerr 1992, p. 451

11. Buffon, translated by Smellie, 1781
12. *Ibid.*, p. 34
13. Whitehurst 1786, Preface
14. Furniss 2018, p. 131
15. See https://www.encyclopedia.com/science/dictionaries-thesauruses-pictures-and-press-releases/toulmin-george-hoggart
16. See, for example, McIntyre 1963 and Davies 1967
17. Porter 1978, *George Hoggart Toulmin's theory of man and the earth*, pp. 343–5
18. Hutton's *Theory of the Earth* 1788, quoted by Porter, 1978
19. Porter 1978, *Philosophy and Politics of a Geologist*, p. 439

CHAPTER 14

1. Walker to Lord Kames NRS GD24/I/571/164-170, quoted in Eddy 2003, p. 135
2. Hutton 1788, p209. Quotations are from *Transactions of the Royal Society of Edinburgh*, Vol. 1
3. The expression 'uniformitarianism' is generally attributed to William Whewell, who introduced the term in 1832. It was widely used by Lyell.
4. See Perrin 1983
5. Dean 1992, p. 11
6. Playfair 1803, pp. 22–3

CHAPTER 15

1. Quoted in Dean 1975 and Dean 1992. The original is in the Fitzwilliam Museum, Cambridge.
2. Kölbl-Ebert 2009, p. 84; see also Smitten 2016
3. Dean 1975
4. Dean 1992, pp. 275–6
5. Hutton 1788, p. 209. Quotations are from *Transactions of the Royal Society of Edinburgh*, Vol 1
6. Merolle 1995, Vol. II, p. 332
7. BL MS 36297 A f3

8. Black to Dashkova (Edinburgh, 27 August 1787)

9. Dean 1992, p. 56

10. *Ibid.*

11. Williams 1789, p. xxv

12. *Ibid.*, p. lxi

13. *Monthly Review* v6 1791, https://babel.hathitrust.org/cgi/pt?id=hvd. hxjg9n&view=1up&seq=138

14. Playfair 1803, pp. 47–8

15. De Luc 1791, p. 138

16. Rudwick 2007, p. 333

17. Kirwan 1793, p. 64

CHAPTER 16

1. Walker 1812, p. 339

2. Hall 1805, p. 74

3. Hall, Sir James, of Dunglass (1761–1832), Jean Jones, Oxford DNB https://doi.org/10.1093/ref:odnb/11965

4. Hall read his paper on February 4, March 3 and April 7 1788, 'but did not incline that this paper, or any abstract of it should be printed'. *Transactions of the Royal Society of Edinburgh*, Vol. 2, 1790

5. See Allchin 1994

6. Donovan 1978, p. 178

7. Dean 1992, p. 31, but Hutton did not read it to the Royal Society of Edinburgh until 1790–1, *Transactions of the Royal Society of Edinburgh* vol 3(2), pp. 77–85

8. McIntyre 1997, p. 113

9. Craig, McIntyre & Waterston 1978

10. McIntyre 1997, p. 141

11. Proceedings of the Royal Society of London, Vol. 3, p. 10

12. Hutton 1899, Vol. III, Ch. IV, pp. 16–17

13. *Transactions of the Royal Society of Edinburgh* vol. 3(2), p. 79

14. *Ibid.*, p. 80

15. Tomkeieff 1953

16. Hutton 1899, Vol. III, p. 232

17. *Ibid.*, p. 236

18. Hutton, 1795, Vol. I, p. 473

19. *Ibid.*, p. 497

20. *Ibid.*, p. 501
21. Playfair 1803, pp. 34–5
22. Hutton to Watt, 23 Sep 1788, Jones, Torrens & Robinson, 1995, p. 371

CHAPTER 17

1. Rae 1895, p. 432
2. Drescher 1999, p. 214
3. *Ibid.*, p. 434
4. Hutton, James: 'A note on Adam Smith's death', *Transactions of the Royal Society of Edinburgh,* Vol. 3, p. 131
5. Cockburn 1977, p. 45
6. Playfair 1803, p. 59
7. Black, Joseph and Hutton, James, 1795, *Essays on Philosophical Subjects by the late Adam Smith LLD*, Cadell, Davies and Creech
8. Smith, *Lectures on Rhetoric and Belles Lettres*, quoted in the Glasgow Edition of the *Works and Correspondence of Adam Smith (1981–87)* Vol. III Es, p. 6
9. https://www.rcoa.ac.uk/about-college/heritage/history-anaesthesia
10. Russell became a Fellow of the Incorporation of Surgeons in 1777, a year before it became the Royal College of Surgeons of Edinburgh. In 1796 he was elected President of the College.
11. Black to Watt, 1 Dec 1791, Jones, Torrens & Robinson 1995, p. 186
12. James Hutton to Alexander Smellie. 1792, NMS MS 592, No. 1
13. Hutton 1792
14. Playfair 1803, p. 40
15. 22 Jan 1975, University of Edinburgh special collections, La.II.646/128.1
16. Hutton 1794 (1), p. 42
17. Greenfield to Hutton, 1 Sep 1794, BL Add MS 36297 A. Greenfield was disgraced and forced to resign over a scandal, but this was after Hutton's death.
18. ($CaCO_3 \rightarrow CaO + CO_2$), see Donovan 1978 and Gerstner 1968
19. Gerstner 1968
20. Gerstner 1971, p. 353
21. *Ibid.*, p. 355

22. *British Critic*, Vol. 7 1796, p. 352

23. *Monthly Review*. ser.2: v.16, 1795, p. 246

24. Gerstner 1971, p. 361

CHAPTER 18

1. Jones, Peter, 1984, p. 182

2. McIntyre 1997, p. 118

3. Robison to Watt, 25 Feb 1800, Jones, Torrens & Robinson 1995, p. 337

4. Allchin 1994, p. 619; Hutton 1792, p. 184

5. Hutton, *ibid.*

6. Playfair 1803, p. 39

7. Playfair 1803, p. 40

8. Jones, Peter, 1984, p. 184

9. NLS MS.5368, f.76

10. Hutton 1794 (2) Vol. 1, p. *i*

11. *Ibid.*, p. ii

12. *Ibid.*, p. 469

13. O'Rourke 1978, p. 19

14. A proposition first advanced in Porter 1977.

15. Gould 1987, pp. 61–96

16. Torrens papers, p. 27

17. See Jones 1985 and Jones, Torrens & Robinson, 1995, pp. 374, 380

18. Watt to Hutton, 22 Dec 1795, Jones, Torrens & Robinson, 1995, p. 380

19. *Transactions of the Royal Society of Edinburgh* vol. 6, pp. 71–175

20. *Ibid.*, p175. Hall made strenuous efforts to confirm with specific experimental results the conclusions Hutton had arrived at through observation. He also did experiments on melting rock and is widely regarded as the 'Father of Experimental Geology'. See Chapter 21.

21. Hutton 1794 (2), Vol. 2, p. 314

22. *Ibid.*, p. 252

23. *Ibid.*, p. 180

24. *Ibid.*, p. 272

25. *Ibid.*, p. 303

26. *Ibid.*, p. 304

27. *Ibid.*, p. 309
28. *Ibid.*, p. 310
29. *Ibid.*, p. 337
30. *Ibid.*,
31. Peter Jones (1984) classifies the three volumes as the nature of ideas, the nature of reasoning and the nature of morality.
32. Hutton 1794 (2), Vol. 3, p. 539
33. *Ibid.*, p. 543
34. *Ibid.*, p. 586
35. https://www.nrscotland.gov.uk/research/guides/slavery-and-the-slave-trade
36. Hair 2000, p. 145
37. See Duckham 1969, p. 186
38. *Ibid.*, p. 194
39. In an undated draft letter, John Clerk complains that coal output from his mine is down by a third because of the 'mutinous' attitude of the colliers since their emancipation and his neglect of the management, necessitating borrowing from 'our most benevolent and worthy friend Dr Hutton'. He worries that in order to pay this and other debts he might have to sell his house in Princes Street, Edinburgh. NRS GD18/5486/53/1
40. Book III, Ch. II
41. Hutton 1794 (2), Vol. 3, p. 555
42. *Ibid.*, p. 587
43. Hutton to George Clark, July 1755, NRS GD18/574

CHAPTER 19

1. *English Review, Or, An Abstract of English and Foreign Literature, Curiosities and Wonders,* Vol. 24, J. Murray, 1795, p. 431
2. *Critical Review: Or, Annals of Literature,* Vol. 19, Baldwin, 1797, p. 308
3. *Analytical Review, Or History of Literature, Domestic and Foreign, on an Enlarged Plan,* Vol. 21 J. Johnson, 1795, p. 1
4. Black to Watt, 16 Jan 1794, Robinson & McKie 1970, p. 197
5. Black to Watt, 6 Jun 1794, Robinson & McKie 1970, p. 202
6. Black to Watt, 12 Apr 1795, Robinson & McKie 1970, p. 215
7. Dean 1992, p. 62
8. Watt to Hutton, 9 Dec 1795, Jones, Torrens & Robinson, 1995, p. 379

9. Dean 1992, p. 62
10. For a detailed list of the changes and additions, see Dean 1992, p. 66–7
11. Hutton 1795, Vol. I, p. 201
12. *Ibid.*, p. 204
13. *Ibid.*, p. 215
14. *Ibid.*, p. 222
15. *Ibid.*, p. 240
16. *Ibid.*, p. 369
17. Hutton 1795, Vol. II, p. 545
18. *Ibid.*, p. 547
19. Dean 1992, p. 78
20. *British Critic*, Vol. 8, 1796, Hathi Trust Digital Library, p. 606
21. *Transactions of the Royal Irish Academy*, Vol. 6, p. 234
22. Howard 1797, p. 549
23. *British Critic*, Vol. 9, 1797

CHAPTER 20

1. Playfair 1803, p. 45
2. Hutton 1795, Vol. II, p. 567
3. Geikie, preface to Hutton 1899
4. Jean Jones dated sections of the manuscript from watermarks on the paper, see her handwritten notes Torrens papers p. 80
5. Hutton, *Elements*, p. 10
6. Jones 1985, p. 575
7. Hutton *Elements*, p. 500, quoted in Pearson
8. Ingenhousz 1779
9. Hutton *Elements*, Section 4, Ch. 7, pp. 309–10. I am grateful to Professor Alan Werritty, who is preparing Hutton's ms for publication, for drawing my attention to this and other sections.
10. Playfair 1803, p. 50
11. Robison to Watt, Edinburgh 7 April 1797, quoted in Muirhead 1854
12. *The Edinburgh Advertiser*, 31 March 1797
13. 28/03/1797, Old Parish Registers Deaths 685/1 980 281 Edinburgh, p. 281. 'Keiss' may be a mistranscription for 'Kerr'.
14. It was refurbished in 2017 by the James Hutton Institute.

15. Black to Watt, 13 June 1797, quoted in Muirhead 1854
16. NRS MFilP/C22/91/33
17. NRS RD5/132 pp. 181–229: I am grateful to Anne Pack and Jeanne Donovan for allowing me to use their transcription of Isabella's will.
18. NRS MfilP/C22/91/33
19. Post Office directory for 1820–21

Chapter 21

1. Playfair 1803, p. 55
2. Torrens papers, pp. 505–9
3. NLS Acc10000/12 p. 13
4. For a fuller explanation of these terms and others see Dean 1992, pp. 95–6
5. Jones 1984
6. Jones 1985, p. 575
7. Cockburn 1977, p. 182
8. Horner's note accompanying the ms is reproduced in Dean 1997
9. Bonney, T.G. 1899, 'Theory of the Earth with Proofs and Illustrations', *Nature* 60, p. 220. https://doi.org/10.1038/060220a0
10. Craig, McIntyre & Waterston 1978
11. Bonney, *op. cit.*
12. Cockburn, p. 51
13. Hope developed his paper into a book, *An Account of a mineral from Strontian* (1798), and inscribed an author's copy to Hutton. However, it arrived after his death. It is now in the library of the University of St Andrews.
14. Porter 1977, p. 145
15. Dean 1992, p. 85
16. *Transactions of the Royal Society of Edinburgh*, vol. 5, pp. 43–75
17. *Transactions of the Royal Society of Edinburgh*, vol. 3, p. 8
18. *Transactions of the Royal Irish Academy*, vol. 8, pp. 21–7
19. Dean 1992, p. 104
20. Playfair 1802
21. Dean 1992, p. 116
22. Playfair 1802, p. 181
23. Playfair 1802, pp. 437–40
24. Morrell 1971, pp. 50–2

25. Playfair 1802, p. 125

26. Playfair 1802, p. 139

27. *Edinburgh Review* 1802, p. 202

28. *Ibid.*, p. 215

29. Murray 1802, pp. 252–3

30. *Transactions of the Royal Irish Academy*, vol. 9, pp. 430–1

31. Quoted in Dean 1992, p. 134

32. *Memoirs of the Wernerian Natural History Society*, 1811. United Kingdom: Bell & Bradfute, etc. p. 130

33. See Geikie 1905 and Craig, McIntyre & Waterston 1978, p. 8

34. Flinn 1980, p. 252

35. *Ibid.*, p. 256

36. Thackray 1998, p. 18

37. Dean 1992, pp. 146–9

38. *Ibid.*, p. 169

CHAPTER 22

1. Robert Bakewell the geologist (1768–1843) was not related to the agriculturalist of the same name, who was quoted by Hutton in *Elements of Agriculture*.

2. *Ibid.*, p. 138

3. Bakewell 1833, p. 135

4. Wilson 1998, p. 23

5. Thackray 1998, pp. 18–19

6. Lyell 1830

7. *Ibid.*, p. 4

8. Wilson 1998, p. 26

9. Thackray 1998, pp. 18–19

10. *Edinburgh Review*, 69, 1839, p. 406

11. *Ibid.*, p. 411

12. Lyell, Mrs, 1881, *Life, Letters and Journals of Sir Charles Lyell, Bart*, Vol, 2, p. 47, quoted in Eyles 1947

13. Thomson 1868, para 19

14. *Ibid.*, para 26

15. Rognstad

16. Geikie 1895, p. 116

17. Geikie 1858, p. 245

18. Geikie 1924, p. 308
19. Geikie 1899, p. xiv
20. Geikie 1905, *passim.*
21. Dean 1992, p. 261. From geological evidence, Geikie estimated the earth to be 73–680 million years old.
22. Perry 1898
23. Quoted in England *et al.* They argue that it was Perry's intervention which was decisive and that the new source of heat from radiation would have made no difference to Thomson's calculations.

CHAPTER 23

1. Geikie 1899, p. xvi
2. The Geological Society of London and the Royal Society of Edinburgh, in association with INHIGEO, held a conference on the 200th anniversary of Hutton's death, which coincided with that of the birth of Lyell.
3. Tomkeieff 1947, p. 253
4. *Ibid.*, p. 256
5. *Ibid.*, p. 258
6. Gohau 1997
7. Tomkeieff 1962
8. Dean 1992, p. 215
9. NLS MS.5368, f.76, James Hutton to Andrew Stuart, undated but probably first quarter 1773
10. Repcheck 2003
11. Baxter 2003

Index